Research and Development of
Gas Turbine Performance

Research and Development of Gas Turbine Performance

Edited by **Eugene Bradley**

New York

Published by NY Research Press,
23 West, 55th Street, Suite 816,
New York, NY 10019, USA
www.nyresearchpress.com

Research and Development of Gas Turbine Performance
Edited by Eugene Bradley

International Standard Book Number: 978-1-63238-400-3 (Hardback)

Printed in the United States of America.

Contents

Preface

This book discusses research-focused information regarding the development of gas turbine performance. There has been a remarkable transformation in the research and advancement regarding gas turbine technology for transportation and power generation. The former remains unchanged with respect to the past, as the superiority of air-breathing engines in comparison to other technologies is immense. On the other hand, the world of gas turbines (GTs) for power generation is characterized by various scenarios, as new challenges are coming up in current energy trends, where the reduction in the usage of carbon-based fuels and the increase in the utilization of renewable sources are becoming highly significant. It has been a key technology for base-load applications for many years; however major stationary gas turbines are facing challenges in balancing electricity from various renewable sources with that from flexible traditional power plants. The book presents an updated picture as well as a perspective view of some of the above mentioned topics that distinguish GT technology in two different applications- aircraft propulsion and stationary power generation. So, the target audience for the book includes design, analyst, elements and maintenance engineers. Some major topics included within the book are gas turbine and element performance, gas turbine combustion, and fault revelation in systems and substances.

Various studies have approached the subject by analyzing it with a single perspective, but the present book provides diverse methodologies and techniques to address this field. This book contains theories and applications needed for understanding the subject from different perspectives. The aim is to keep the readers informed about the progress in the field; therefore, the contributions were carefully examined to compile novel researches by specialists from across the globe.

Indeed, the job of the editor is the most crucial and challenging in compiling all chapters into a single book. In the end, I would extend my sincere thanks to the chapter authors for their profound work. I am also thankful for the support provided by my family and colleagues during the compilation of this book.

Editor

Gas Turbine and Component Performance

Synthesis of Flow Simulation Methods for Fast and Accurate Gas Turbine Engine Performance Estimation

Ioannis Templalexis

Additional information is available at the end of the chapter

1. Introduction

Gas turbine engine development and maintenance comprises a great amount of risk. Nowadays a company on its own, the engine manufacturer for instance, cannot afford the entire engine development risk. At the same time, manufacturers need to provide their customers, the engine users, with a competitive maintenance package. Investments on gas turbine engine development and maintenance were magnified over the years, a process driven both by the competition and strict airworthiness and environmental regulations. Consequently, manufacturers are looking both for risk share partners and cost shrinkage.

A principal tool to achieve the aforementioned goals is computer based, engine performance simulation. Risk share partners need to have a view, or better to say, an evidence regarding the performance of the engine under development. Gas turbine engine performance simulation however, has a much greater impact on narrowing down the engine development related cost and on providing early evidence of engine malfunction, thus suppressing also the maintenance costs.

Computer based gas turbine engine performance simulation and the derived methods are classified and selected for a particular application, based on the leverage between accuracy and computational load. On one end of the classification scale stand the zero dimensional (0-D) methods and on the opposite end stand the so called Computational Fluid Dynamics (CFD) methods based on Reynolds Averaged Navier Stokes (RANS) equations.

The current chapter aims to present a gas turbine engine, tailor made, performance simulation tool that stands out as an optimum combination of accuracy and execution cost. The cost of applying a certain simulation method rises with computational load. The architecture of the simulation tool under context is justified and at the same time it takes advantage of the fact that the prescribed accuracy for the simulation of each component is not the same. Conse-quently a single flow simulation tool, when applied on the entire engine would eventually cover the prescribed requirements for simulation of certain components but not of the entire engine. The simulation methodology that is to be presented over the following paragraphs, considers variable dimensions regarding the flow resolution of each component in order to address firmly gas turbine engine simulation requirements.

The current chapter, consists of two major sections. The first one addresses the simulation tools and methods for each component. The second section describes the amalgamation of these methods and tools to give the integrated gas turbine engine simulation tool. Before that a brief literature review will be given regarding similar efforts.

2. Literature review

Gas turbine engine simulation has been addressed as an issue by several researchers during the past. The current section aims to list an indicative sample of relevant research work and associated software. The author will not go into details, however readers that are further interested on this specific research field, are advised to address to the citations given below. Joachim Kurzke [1], has been very active on gas turbine engine simulation, specialized on 0-D flow simulation and component map generation. Also M.G. Turner et al [2] have dealt with gas turbine engine simulation, producing hybrid tools aiming to reduce computational load, while retaining an acceptable accuracy level. Alexiou et al [3], [4], have also been very active in the research field of engine flow simulation Additionally Alan Hale et al [5], have developed hybrid tools for effective gas turbine engine simulation. Finally the author of the current chapter working in conjunction with Cranfield University research team and in particular with Dr. Vasilios Pachidis has contributed substantially to the research field under context [6], [7], [8]. Very little gas turbine engine simulation software has been developed up to the moment capable of capturing the engine performance alternation under uneven inlet flow conditions. Several flow calculation methods are available in the open literature, but the computational resources required, for the system of equations to be solved numerically throughout the gas path, are enormous. Such a code is the main outcome from the Numerical Propulsion System Simulations (NPSS) project realized at the NASA Glenn Research Center (GRC). A high fidelity 3-D gas turbine performance code has been developed for gas turbine engine simulation. In NASA report [9] a case study of the CF-90 high by-pass ratio is presented.

3. Fields of application for gas turbine engine simulation tools

As it was mentioned in the introductory section, there exists a very wide range of flow simulation packages that are tailored to, or can be applied for, gas turbine engine performance simulation. A vast categorization of these methods and derived tools, is usually made on the basis of the number of dimensions considered for the flow analysis. Alternatively we very often see those methods to be categorized according to the level of flow complexity considered. In specific, according to the former categorization we have:

i.   Zero Dimensional (0-D)

ii.   One Dimensional (1-D)

iii.   Two Dimensional (2-D)

iv.   Three Dimensional (3-D)

according to the later categorization we have:

i.   RANS

ii.   Euler based

iii.   Potential flow based

Flow simulation methods and tools out of all the aforementioned categories could potentially be applied along a certain sector of the engine development road. On figure 1 the range of application for a certain category, is shown as a vast proportion of the entire development road of the engine, starting from the initial specifications up to the engine testing. Ideally the 3-D high fidelity methods could be applied during the entire development period of the engine. However their application range as it can be seen on Fig. 1 is restricted to the right end. That is just before the final manufacturing drawings are sent for metal cutting. The increased simulation fidelity of these tools comes with an increased investment and operational cost on computational power. Several hours or even days in some cases stand in between setting the inputs and obtaining a converged solution.

It remains questionable whether the increasingly available computational power at a lower cost will shrink the convergence time since more and more sophisticated turbulence and viscosity models appear to give a more accurate flow solution at the expense of the excess computational power available. As a result of the above, the usage of high fidelity 3-D flow simulation tools can be economically justified when applied to the final component or engine design. The aim is to trim the final design in order to match the closest possible the pre-defined performance. What should be avoided at all costs is to be obliged to re-design and re-fabricate a part or the entire component as a result of the fact that the experimental results do not match to the prescribed requirements. Consequently high fidelity simulation is welcomed, but it should only be used when such high fidelity results are absolutely necessary.

On the other hand, focusing on the left end (always according to Fig. 1) a narrow application range of 0-D and 1-D methods is spotted. That is actually during the preliminary design phase, when a wide range of a large number of design parameters are still under discussion resulting in a large amount of simulation cases to be examined. At this stage, qualitative assessment is of primary interest as opposed to quantitative assessment that becomes increasingly important during the following development phases. Low fidelity – low computational load methods match perfectly to the requirements of the preliminary design phase.

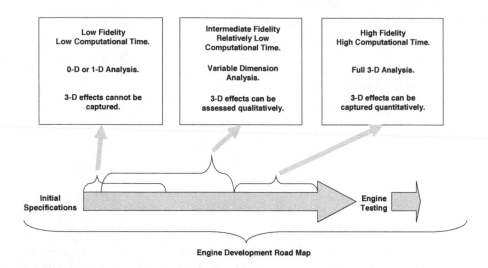

Figure 1. Field of Application for Gas Turbine Engine Simulation Tools

In between low fidelity and high fidelity methods and derived tools, stand the hybrid flow simulation methods that could be identified as intermediate fidelity – relatively low computational time methods. Attempting to underline the identifying characteristic of these methods one would conclude that the required computational time before convergence is not an important issue to consider. In other words the number of case studies is not determined by computational load restrictions. Requiring the minimum possible computational time consumption while demanding an increased fidelity of the derived results, those methods end up having the following characteristics:

i.   They usually are 2-D methods, treating the flow as in-viscid. A 3-D method would be accepted when solving the flow as in-viscid and irrotational.

ii.   Component performance disturbances due to 2-D and even 3-D non-uniformities are at a certain accuracy level captured and assessed through hybrid simulation tools applying different simulation methods between components.

iii.   Performance information encoded in experimental results or in CFD results is adhered by these simulation tools in a tabular format (Component performance map) or even better in the form of empirical equations.

Based on the above mentioned characteristics of these methods and as far as their application range is concerned, intermediate fidelity methods are continuously expanding towards both ends of the engine development trail. It is not surprising that nowadays these methods are gaining increasing attention by engine manufacturers and research institutes. It is the author's view that given the cost reduction of computational power such methods will completely cover the range of applications of 1-D and 0-D methods. The fraction of the application field of the high fidelity methods that can be covered by these methods depends on the ability of a specific tool to adopt experimental results.

4. Components of the hybrid simulation method

The current section explicitly refers to the components of the hybrid simulation method. More specifically it refers:

1. To the method and derived tools for gas turbine engine intake simulation, where the flow is considered as potential. The simulation tool is based on the Vortex Lattice Method (VLM).

2. To the methods and derived tools for the flow simulation of the most sensitive engine component, the compressor. The two modules of the proposed flow simulation strategy, are based on "multi parallel compressor method" and " Streamline Curvature Method" respectively.

3. To the method and derived tools for the flow simulation of the remaining engine components (Combustion chamber, turbine, nozzle). The performance simulation tool is based on the 0-D performance simulation method.

The architecture of the described Intake – Gas Turbine Variable Dimension Performance Simulation Method, will be explained during the second main section of the current chapter. It will then be more clear to the reader the reasons for selecting a specific method for each component as detailed to the three points above.

4.1. Intake section

4.1.1. Intake flow simulation method

The flow within the engine intake is treated as in-viscid. The three dimensional flow non-uniformities are to be carried through the intake up to the compressor face, while keeping the computational load to the minimum possible. A flow simulation method that meets the above mentioned pair of requirements is the VLM. The VLM was among the earliest of such methods written in computer code, mostly addressed for the computation of airfoil aerodynamic characteristics. It was first conceived at the late 30's, but it could not be applied efficiently until the early 60's. The reason was that being a purely numerical method involving big matrix inversions, it had to wait for the computers to develop sufficiently in order to support such calculation load [10].

It belongs as a method to the group of Panel Methods (PM) since:

• The linear potential flow equation is solved.

• Panels are used for the description of the geometry.

• Singularities are placed on a surface.

- The "Neumann" boundary condition is applied to a number of control points.

- A system of algebraic equations is solved to determine the singularities strength.

What makes the VLM a separate method are the facts that:

- It is usually applied on lifting surfaces.

- One kind of singularity (vortex filaments) is strictly used. It is not distributed over the entire surface, but only along the surface boundaries.

According to the VLM the flow field around a lifting surface is established by superimposing the free stream flow velocity to the velocities induced by the vortex filaments. It is reminded that vortex filaments have constant circulation Γ along a certain vertex line which cannot begin or end abruptly in a fluid. It must either be closed, extend to infinity or end at a solid boundary. The circulation about any section is the vortex strength. A vortex obeys to the Biot-Savart law and according to its specific shape it has a specific mathematical expression for expressing the induced velocity, at any arbitrary point in space. [11]. It is reminded that the Biot-Savart law was initially defined for the description of the magnetic flow field induced around a current carrying conductor. It is also used for the determination of the velocity vector field appearing around a vortex filament, due to the obvious similarity between the two physical phenomena. For a general 3-D vortex filament (see Figure 2) the velocity increment induced at an arbitrary point in space (P), by an infinitesimal vortex segment (dl) is given by the following equation:

$$dV_p = \frac{\Gamma}{4 \cdot \pi} \cdot \frac{dl \times r_{dl,p}}{|r_{dl,p}|^3} \tag{1}$$

Where Γ the circulation around the filament (constant).

Hence the velocity at point P due to the entire vortex filament would be:

$$V_p = \frac{\Gamma}{4 \cdot \pi} \cdot \int \frac{dl \times r_{dl,r}}{|r_{dl,r}|^3} \tag{2}$$

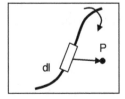

Figure 2. Vortex filament.

The evaluation of the above integral is not always a simple straightforward calculation procedure. The vortex filaments that are used as "construction elements" for the VLM have a well defined and documented solution as they have been used extensively during the past. Vortex filament types can be combined to give new singularity elements ready to be used by

VLM calculation schemes, just like the source and the sink when combined together give a doublet. The most widely used "synthesized" vortex is the horseshoe vortex. It consists of three straight line vortices. One finite length vortex and two semi-infinite vortices. It is basically the model of a finite wing. The two semi-infinite vortices represent the wing tip vortices and the finite vortex represents the wing span. According to the literature review that has been done the numerical schemes using purely vortex singularities can all be classified into two major categories. In the first category horseshoe kind of vortex filaments are incorporated and in the second category the closed bound vortex filaments around the panel surfaces are used.

A horseshoe vortex method would not be attractive for the intake internal flow simulation. Consequently the only candidate left is the "closed vortex filaments" method [11], having more advantages when compared to the rest of the flow simulation methods, for the development of a rapid calculation scheme to be applied in the intake internal flow regime. Both the flow around lifting and non lifting surfaces can be equally well simulated. Various alternative solutions for the wake handling can be incorporated while very little computer resources are required for convergence.

In particular the VLM used for the intake flow simulation is summarized by the following application steps. Firstly the geometry is defined using an arbitrary number of panels in the sense that there is not any upper limit other than the computer processor capabilities. A sensitivity analysis concerning the optimum number of panels to be used fits perfectly at this point. The panels are flat of triangular or rectangular shape. Their control points coincide with their geometrical centers, where a local Cartesian coordinate system is being defined by a normal and two tangential vectors. Secondly a closed vortex filament distribution is assumed over the surface surrounding the panel boundaries. The "Neumann" boundary condition is applied resulting in a linear system of n x n independent equations, where n is the number of panels. The solution of this system gives the vortex strength matrix. The Kutta – Jukofski boundary condition is applied at the panel trailing edges where wake is expected to be developed. The wake is left free to relax according to the induced velocities. The flow vector field is continuously updated through the iterative process, by the velocity components induced from the wake panels generated. The convergence criterion is based on the aerodynamic forces induced on the panels.

4.1.2. Intake flow simulation software description

The intake flow simulation software, as any 3-D flow simulation software consists of the three following main modules:

- Pre-possessing which encounters the geometry definition and the solution grid settings where applicable.

- Processing, or in other words the flow solver. The module that derives the mathematical solution.

- Post-possessing, which refers to the way the developer chooses to present the results calculated by the above mentioned module.

Developers of 3D flow simulation software, despite the fact they think in terms of the three distinct modules mentioned above, these modules are not always distinguishable by the end user. The pre-processing module should ask and acquire from the user in a certain numerical format the set of geometrical data that lead to the geometry related to the flow simulation regime (it is irrelevant whether intermediate calculations take place). The pre-processor of the VLM tool under context is used to create the intake geometry covered in panels. It is one of the input files fed into the processing unit.

The processor receives the input files, conducts the calculations and delivers the set of output files for post processing. It is the core of a CFD software. An integral solver does not contain only the set of equations ready to be solved, but also a set of subroutines in order to reassure convergence, under any set of inputs. Additionally the solver should ideally inform and consult the user during execution.

The post processing of the results is not necessarily done, although it is very convenient, by the same software that conducts the pre-processing is also highly dependent upon the needs of a case study. The VLM software as a post-processor offers only the visualization of stream-lines.

4.2. Compressor section

4.2.1. Compressor flow simulation method

Attempting to classify gas turbine engine components based on the flow regime in terms of complexity, temperature and pressure, three major groups can be recognized: The components located upstream of the combustion chamber, the components located downstream of the combustion chamber and the combustion chamber itself. Within the "upstream" components, the pressure is rising in the direction of the flow, whereas the opposite takes place along the "downstream" components. Upstream of the combustion chamber due to the "un-physical" character of the process, since the air is forced to move against an adverse pressure gradient, the flow is becoming very complex and often unpredictable. This is even more pronounced in the compressor, as it consists of both rotating and non-rotating parts. The flow within the combustion chamber is extremely complex, as it is actually controlled by the thermodynamics and chemistry of the combustion process. Finally, within the working medium expansion region of the engine, the flow can be considered relatively simple from the fluid mechanics point of view.

It is therefore evident that the most critical component of the gas turbine engine regarding its operation and its response to the changing operating conditions is the compressor. Conse-quently the compressor becomes very challenging, when it comes to the flow simulation under the requirement for minimum computational power consumption. Flow non-uniformities conveyed through the engine intake down to the engine face affect severely the compressor overall performance and operational stability. The later refers to the minimum flow that can be handled before surge occurs. Therefore the compressor response should be known through-out its operational envelope, for all possible inlet conditions and throttle schedules. Using a single flow simulation method to assess the compressor performance would unavoidably lead to a 3-D tool. Euler or RANS based methods are excluded when referring to fast assessment tools. On the other hand VLM cannot cope with the flow regime present in a compressor as it is highly turbulent and 3-D boundary layers are developed especially at the rear stages. Additionally the application of VLM method in the case of a multistage compressor simulation demonstrates several geometry and convergence handling problems, where the wake handling problem stands above all. This problem arises only in the case where one blade row is immediately followed by a second one (rotor stator). Even in the case where the stator wake passes smoothly through the rotor during the first iterations without causing any instabilities, a few iterations later, when the rotor would have rotated by 60 degrees for example, the wake twists un-physically, remaining trapped between the rotor blades it first crossed. The physical process dictates that the following row truncates the wake generated by the blades of the upstream blade row. Attempting to program the solver to do so, the phenomenon cannot still be simulated very accurately because it is impossible for the wake to be cut down to less than one panel. The axial distance between blade rows in modern compressors is less than a wake panel's length. On the top of that, wake is dissipated as it travels down between the moving and stationary blades and it is not conserved as in the case of Panel Methods.

Consequently the 3-D flow profile present at the compressor inlet plane has to be de-composed and let each component be treated by a separate tool. In the case presented herein, the profile is decomposed to a radial and a circumferential component. The former is treated with a streamline curvature method based tool and the later is treated with a multiparallel compressor method based tool.

4.2.2. Streamline curvature flow simulation method

The origin of the SLC method, before even being identified as a separate calculation method, lies on Wu's through-flow theory [12]. The method under its current name was developed independently by Smith [13] and Novak [14] in the United States and Silvester and Hetherighon [15] in the United Kingdom. However, at that time the solutions were still restricted to the duct regions. The basic idea lies on the integration of the full radial equilibrium equation across the blade edges in conjunction with the flow continuity equation, for the determination of the meridional velocity profile across the compressor. Frost [16] took the method a step further, by applying it within the blade rows, demonstrating thus the first SLC-method-based flow representation within a compressor. Finally Senoo and Nakese [17] and Novak and Hearsey [18] reported quasi-3D SLC method applications.

During the approximately forty years of existence of the SLC method, numerous authors have proposed several variations of the SLC calculation scheme. All these different schemes were mainly influenced by the type of the turbomachine the method was applied to (radial or axial), the nature of the flow (subsonic or supersonic) being considered and the level up to which the flow viscosity and circumferential in-homogeneities were taken into account. In many cases, the various calculation schemes were also influenced by the particular characteristics of the cascade, such as hub to tip ratio, lean and sweep angle distributions, etc.

During the many years of development of the SLC method, several excursions from the initially proposed solution scheme were made in many aspects. Firstly, differences can be noticed concerning the final form of the SLC equation steaming from the coordinate system selection and the quantities each researcher includes in the solution scheme. For instance, Frost [16] defined a solution grid formed by calculating planes parallel to the upstream and downstream boundaries, whereas Jennious and Stow [19] defined a solution grid by approximating streamlines using a curve fit through points of equal mass flow. Differences can be spotted even in the way the SLC is calculated. Two different approaches are suggested by Shaalan and Dareshyar [20] and Wilkinson [21].

As mentioned previously, the quantities each researcher includes in the final equation have also an impact on the type of the final differential equation. This particular issue is discussed in detail in the third section of this manuscript. It is worth noticing that almost none of the final SLC equations met in various literature texts are the same, although they do have some similarities. Moreover, differences between researchers applying the same method can be found on the solution process depending on whether inter-blade stations are considered and also depending on empirical rules applied to aid the solution process and for controlling the convergence procedure. To cite an example, Casey [22] has considered inter-blade stations, whereas Denton [23] in his very early work did not. Concerning the convergence control, Wilkinson [21] proposed a very interesting study. Finally, another field where different approaches are met is the way each researcher chooses to take into account the presence of viscous forces into the flow, as well as the circumferential variations along the blade span. In some earlier SLC methods those effects were not considered at all. Horlock [24] suggested the inclusion of a non-conservative body force acting opposite to the stream direction in order to make the momentum equation consistent to the in-viscid assumption. Concerning the inclusion of circumferential effects, which are not present in a 2-D method such as the SLC, three models most commonly used have been reviewed by Horlock and March [25]. The models are based on the replacement of the actual cascade with a cascade containing an infinite number of blades, simulating thus the blade action by an axi-symmetric flow with distributed body forces along the blade and by considering the flow on a suitably defined "mean stream surface" [22].

Authors have adopted the 'system approach' in order to derive the REE, for an elementary fluid element moving through the blade rows. The cylindrical coordinates are more appropriate for this study, than any other system of coordinates, given the geometry of a turbomachine. More precisely, the principal directions will be the circumferential, or whirl direction "w", the meridional direction "m" and the normal to the meridional direction "n". The m-n directions are used instead of the axial-radial, because the basic concept of the current method of solution is based on the SLC. In figure 3 the principal and some auxiliary directions are being defined, as well as the angles involved in the derivation process.

The application of Newton's second law of motion, for this elementary fluid element, gives the following system of equations, for the three principal directions [26]:

$$\text{m: } V_m \cdot \frac{dV_m}{dm} - \frac{V_w^2}{r} \cdot \sin\varepsilon = -\frac{1}{\varrho} \cdot \frac{dP}{dm} + F_m \tag{3}$$

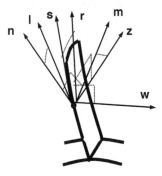

Figure 3. Definition of angles and directions [27]

The first term corresponds to the inertial force due to the acceleration of the element in the meridional direction. The second term denotes the centrifugal force term due to the circumferential movement, projected onto the meridional direction. The third term corresponds to the pressure force in the same direction and the last term represents the body forces exerted onto the element. In the normal direction the equation gives:

$$n: \frac{V_m^2}{r_c} - \frac{V_w^2}{r} \cdot \cos\varepsilon = -\frac{1}{\varrho} \cdot \frac{dP}{dn} + F_n \tag{4}$$

The first term here corresponds to the centrifugal force due to the movement of the element on a curved streamline. The second term denotes the centrifugal force term due to the circumferential movement of the element projected on the normal direction. The third term corresponds to the pressure forces on the same direction and the forth term represents the body forces exerted onto the fluid element. Finally on the circumferential direction Newton's second law gives:

$$w: \frac{V_m}{r} \cdot \frac{d(r \cdot V_w)}{dm} = F_w \tag{5}$$

According to the above equation, the body forces are equal to the corriolis force, since the element moves in the meridional direction, while the whirl velocity component is varying. The above equations are applicable for steady, axi-symmetric in-viscid flow and the final aim is to derive from them a differential equation, the solution of which will give as a result the meridional velocity profile, defined on the m-n plane. Consequently the circumferential flow variations, the flow viscosity and the 3-D nature of the flow, seem to be left out from the analysis. However, this is not entirely true, because some mathematical manipulations were proposed in order to take into account those effects artificially. It is more convenient for the solution process, to have the final equation expressed in s-m coordinates, rather than in n-m coordinates, since the meridional velocity profile is required along the blade leading and trailing edges. The variation of pressure along the s direction is given by the following equation that steams out from the state equation:

$$\frac{1}{\varrho} \cdot \frac{dP}{ds} = \frac{dH}{ds} - T \cdot \frac{dS}{ds} - \frac{V_m \cdot dV_m}{ds} - \frac{V_w \cdot dV_w}{ds} \tag{6}$$

Equations 3 and 4, involve the pressure variation along the meridional and normal directions respectively. Substituting those equations into equation 6, the following equation results:

$$\cos (\varepsilon - \gamma) \cdot \left(F_n - \frac{V_m^2}{r_c} + \frac{V_w^2}{r} \cdot \cos \varepsilon\right) + \sin (\varepsilon - \gamma) \cdot \left(F_m - \frac{V_m \cdot dV_m}{dm} + \frac{V_w^2}{r} \cdot \sin \varepsilon\right) = \frac{dH}{ds} - T \cdot \frac{DS}{ds} - \frac{V_m \cdot dV_m}{ds} - \frac{V_w \cdot dV_w}{ds} \tag{7}$$

Introducing rothalpy into equation 7 instead of enthalpy, a much more convenient quantity when examining rotating flow, and moving from the absolute system of reference to the relative, by substituting:

$$V_w = W_w + \omega \cdot r \tag{8}$$

The following equation appears:

$$\frac{V_m dV_m}{ds} = \sin (\varepsilon - \gamma) \cdot \frac{V_m dV_m}{dm} + \cos (\varepsilon - \gamma) \cdot \frac{V_m^2}{r_c} + \frac{dH}{ds} - T \cdot \frac{dS}{ds} - \sin (\varepsilon - \gamma) \cdot F_m - \cos (\varepsilon - \gamma) \cdot F_n - \frac{V_w}{r} \cdot \frac{d(r \cdot V_w)}{ds} \tag{9}$$

As it was mentioned before, the two main flow effects left out from the equation are the fluid viscosity because the flow is treated as in-viscid and the pressure variation across the blade to blade direction due to the 3-D treatment of the flow. The first effect is represented by a drag force term F_D acting on the opposite direction of the flow and the second effect is taken into account by the introduction of a pressure force acting normal to the side surface of the infinitesimal volume. The drag force is related to the loss mechanism of the flow, thus related to the entropy generation in the meridional direction:

$$F_D = -\cos \beta \cdot T \cdot \frac{ds}{dm} \tag{10}$$

The direction vector of the pressure force according to figure 4 equals:

$$\frac{\vec{F}_P}{F_P} = -\sin \beta \cdot \cos \lambda \cdot \vec{i}_m + \cos \beta \cdot \cos \lambda \cdot \vec{j}_w + \left(\frac{\sin \lambda \cdot \cos \beta}{\cos (-\gamma)} + \cos \lambda \cdot \sin \beta \cdot \tan (\varepsilon - \gamma)\right) \cdot \vec{k}_n \tag{11}$$

However, in equation 9 the force terms appearing are lying on the normal and meridional directions. Consequently those force terms should be expressed with respect to the drag and pressure forces and then substituted back into equation 9. Finally, the later force terms should also be expressed with respect to the flow parameters in order to bring the final equation into a form so that it depends only on velocity components, flow and geometrical angles and thermodynamical parameters. The drag force is already into such a form in equation 10. As far as the pressure force is concerned the following equation is suggested:

$$F_P = \frac{F_w}{\cos \beta \cdot \cos \lambda} + \frac{\sin \beta}{\cos \lambda} \cdot T \cdot \frac{ds}{dm}$$ (12)

Where Fw according to equations 5 and 7 equals:

$$F_w = \frac{V_m}{r} \cdot \left(\frac{d(r \cdot (W_w + \omega \cdot r))}{dm} \right)$$ (13)

$$F_w = \frac{V_m}{r} \cdot \frac{d(r \cdot W_w)}{dm} + 2 \cdot V_m \cdot \omega \cdot \sin \varepsilon$$ (14)

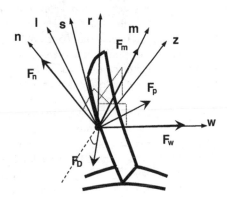

Figure 4. Definition of force vectors [27]

According to figure 4, the force terms F_w and F_p can be expressed with respect to F_m and F_n and then substituted back to equation 9 in order to give:

$$\frac{V_m dV_m}{ds} = \sin(\varepsilon - \gamma) \cdot \frac{V_m dV_m}{dm} + \cos(\varepsilon - \gamma) \cdot \frac{V_m^2}{r_c} - \frac{W_w}{r} \cdot \frac{d(r \cdot W_w)}{ds} - 2 \cdot \omega \cdot W_w \cdot \cos \varepsilon + \frac{dI}{ds} - T \cdot \frac{dS}{ds}$$
$$- \tan \lambda \cdot \left(\frac{V_m}{r} \cdot \frac{d(r \cdot W_w)}{dm} + 2 \cdot V_m \cdot \omega \cdot \sin \lambda \right) + T \cdot \frac{dS}{ds} \cdot ((\sin \varepsilon - \gamma) \cdot \cos^2 \beta - \tan \lambda \cdot \sin \beta \cdot \cos \beta)$$ (15)

The above form of the full REE is found in most relevant texts. It can be used in order to conduct a quick quantitative estimation concerning the flow within turbomachines. The solution of equation 15 will lead to the meridional velocity distribution along the blade leading and trailing edges, or even along inter-blade stations if so defined. However the viscous nature of the flow is not yet fully introduced into calculation. Flow viscosity will generally cause secondary flows and fluid friction against solid surfaces. The aforementioned principal causes will lead to several localized phenomena that are independently studied and quantified on the basis of empirical correlations. There are three major areas of irreversibilities:

i.   Flow deviations on blade leading and trailing edge. The aim of flow deviation models is to define flow incidence and deviation angles in relation to flow and blade profile parameters.

ii.   Boundary layer growth along wetted surfaces. Boundary layer prediction models may end up becoming extremely complex, depending on the level of accuracy one aims to achieve. Care should be taken at this point to balance the computational power consumption against accuracy.

iii.   Frictional losses. Calculating blade losses in compressors is an extremely difficult task due to the complex, three-dimensional nature of the flow field. There are several factors that contribute to the generation of losses in a compressor. It would be fair to say that although consensus exists on the end result, which is an increase in entropy and reduction in total pressure, the exact mechanisms through which losses are generated and their complicated interactions have not been completely understood yet. Several researchers have invested their expertise into correlations of parameters which describe the flow in blade passages. Such correlations usually attempt to synthesize the results of many tests into simpler formulae or sets of curves. They are generally averages of test results or their statistical curve fits. The largest limitation of this approach is that the various empirical correlations can not be expected to sufficiently represent every individual compressor design. A very popular approach to blade row total loss prediction was followed by researchers such as Miller [28] and Creveling and Carmody [29]. It is assumed that the total pressure loss of the blade row is the result of the interaction of different loss components, i.e. profile losses, secondary losses and shock losses which are considered to act through independent mechanisms that is the blade mass flow-averaged total loss factor is given as the sum of specific loss terms:

$$\bar{\omega}_{tot} = \bar{\omega}_{prof} \cdot f_{Re} + \bar{\omega}_{sh} + \bar{\omega}_{sec} \tag{16}$$

4.2.3. Multiparallel compressor method

Circumferential total pressure distortion component is described by dividing the compressor inlet face into a number of "pie" sectors. These sectors are defined by a number of spokes which are intersected at the compressor centerline and they are extended till the compressor outlet casing. An average total pressure value is assigned to each of these sectors.

Multi-parallel compressor method is a common method used to assess the effect of those distortion profile types, on the compressor and gas turbine engine performance. Circumferential total pressure distortion has been a matter of interest for gas turbine engine manufacturers for many years. It was soon realized circumferential total pressure gradients at the compressor inlet face cause degradation on compressor stability margin. This has been recognized to be the most pronounced effect of this type of distortion. However, compressor performance is also affected. In the mid-fifties the effects of circumferential total pressure gradients have been examined experimentally [30], on an axial turbojet engine. This work had been among the first of a kind. Durning the seventies a vast amount of experimental work had been conducted at NASA Lewis research center. In the first report [31], the effect of several screen induced total pressure distortions had been determined on a J-85-GE13 turbojet engine. A very simple modeling technique is developed for the assessment of circumferential total pressure distortion effects on compressor performance. The "Parallel Compressor" model suggested by Pearson and McKenzie [33]. This model had been validated experimentally [32], extended to unsteady versions, [34], [35], [36], included into computer codes (GasTurb) and reviewed by almost any circumferential distortion related paper. Hynes and Greitzer [37] have proposed an alternative modeling method capable of assessing both steady and unsteady flow phenomena, caused by circumferential total pressure gradients. It is an analytical solution method, expressed through a set of differential equations that were derived from conservation laws applied on an integrated compression system consisting of the inlet plenum, the compression system, the outlet plenum and the throttle. Plourde and Stenning [38] had also developed a compressor flow model, for assessing the attenuation of circumferential total pressure gradients, within a multi-stage compressor. This model although not referring directly on the performance degradation, provides useful information and understanding of the flow phenomena induced by this type of distortion.

The current method adopts a parallel compressor model the way that is presented in reference [33], extended to incorporate more than two compressor sectors. The parallel compressor model is formed by dividing the compressor into two compressor segments (Figure 5).

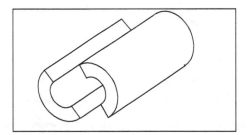

Figure 5. Parallel compressor segments.

Each segment occupies a certain percent of the compressor annulus volume, directly proportional to the angle of extent of each segment. Both segments operate simultaneously and discharge to a common plenum. The first segment is considered to operate under low total inlet pressure and the second is considered to operate under higher inlet total pressure. The fact that both flow regions discharge to a common plenum, justifies the assumption, under certain conditions, that both segments share a common static pressure at their outlet plane. This assumption is valid only if the exit duct is straight and of constant area and the air is leaving the compressor under a uniform stator exit angle. If there is a diffuser or a second compressor downstream of the first compressor, then the flow in that region can no longer be considered as 2-D or the exit static pressure as uniform. Implicit in the above description is that: i) two compressors are working in parallel under different inlet total pressure but at a common exit static pressure, ii) no cross flow occurs within a rotor blade row and iii) no flow redistribution takes place within the axial gaps. This is a rather logical assumption to make, because the rotor tip clearance, especially in modern compressors are rather narrow and sometimes actively controlled by external mechanisms and the axial gaps between successive blade rows are when compared to the circumferential length scales, quite small.

Under the above mentioned assumptions each compressor segment can be considered that it is operating on the same non-dimensional compressor speed line, as the "clean" inlet compressor would operate. The "clean" compressor characteristic for high speed compressors should be given as the pressure and temperature ratio versus the non-dimensional mass flow. Each component segment would not theoretically deviate from the non – dimensional speed line, since the compressor characteristic curves were non – dimensionalized with respect both the inlet total boundary conditions and the geometrical characteristics. Figure 6 shows low and high pressure compressor segment operating points on a typical non-dimensional compressor speed line.

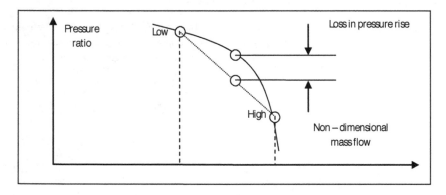

Figure 6. Basic parallel compressor model for compressor response to circumferential total pressure distortion. [39]

Through the parallel compressor model several important aspects of the circumferential inlet total pressure distortion phenomenon can be assessed, at least qualitatively. Firstly as it can be seen on Figure 6, the compressor mean pressure rise is lower than it would have been achieved, if the compressor was operating at the same mean mass flow, uniformly distributed over the compressor inlet face. Also the low inlet total pressure compressor segment produces higher pressure ratio and therefore operates closer to the surge line. Consequently although the compressor average operating point may fall within the stable region of the compressor map, a considerable part of the compressor (the low pressure segment) is working in the unstable region of the map. Consequently flow instability under circumferential total pressure distortion may be initiated earlier (at higher mass flows) than it would have, if the compressor was operating under uniform inlet total pressure. The stall criterion that it is used along with parallel compressor model is that stall occurs when the low pressure region operating point, has crossed the stability line. In order to determine the surge line shift under a given circumferential total pressure profile, the operating point of the low pressure compressor segment, is placed on the surge line. The segment's corresponding mass flow is determined after attenuation by the pie sector angle and subsequently it is non-dimensionilized with respect to the corresponding inlet boundary conditions and cross-sectional area. Consequently the total pressure and temperature rise of the segment are read directly by the "clean" compressor performance maps.

4.2.4. Streamline curvature software description

The SLC based flow simulation software obeys to the architecture mentioned in paragraph 5.1.2. The pre-processing unit having received all necessary geometrical inputs, defines the compressor geometry and sets the initial position of the streamlines. The processing unit upon convergence, will have defined the flow-field and the distribution of all thermodynamic parameters through the compressor operating range. Given the nature of SLC method calculations being highly iterative, the processor incorporates very sophisticated convergence guidance subroutines. Also several alternatives regarding the consideration of viscous phenomena are offered to the user. However it is beyond the scope of this chapter to describe the details of loss models and convergence schemes. There are several publications that the reader could refer to, shall he/she is further interested in the specific field.

The post processing module of a through flow software such as the SLC, would ideally include the streamline visualization through the compressor and preferably the performance map. Options for comparative performance graphs under various inlet conditions and several different geometries turn out to be very informative and useful especially for educational purposes.

The stand alone streamline curvature software SOCRATES has evolved over the past years to handle transonic flow regimes and chocking conditions. A graphical user interface is also under development giving great flexibility to the user regarding compressor geometry definition, loss model selection etc.

4.2.5. Extended compressor in parallel" software description

The "extended compressor in parallel" model is not programmed as an entirely separate code, since it is a straight forward, simple calculation. It is embedded as a subroutine into the 0-D gas turbine engine performance code presented in the following section. The input data is: i) the number of compressor inlet face segments, ii) the angle of extent of each segment, iii) the average inlet total pressure and temperature of each segment and iv) the compressor inlet and outlet cross-sectional area of each segment.

The processing unit houses the calculation procedure which is completed into three concentric iterative loops. The outer most calculation loop is there to repeat the calculations for all possible settings of the variable inlet guide and stator vane angles. The second iterative loop is set to repeat the calculations performed for a single compressor speed line and the third loop is set to establish the exit static pressure balance for all the compressor segments specified in the input file. Finally there is not much post – processing that can be done to the results, other than demonstrating the dislocation of the surge line depending on the inlet distorted temperature and pressure profile.

4.3. Hot engine section

4.3.1. Zero dimensional analysis

The 0-D analysis is an integral type of analysis in the sense that the individual engine components are considered as "black boxes'. Details of the flow within the engine components do not influence the result. Prerequisite for the application of the method are the engine components' performance, in terms of non-dimensional performance maps or alternatively by means of empirical correlations, depending on the information available. The main output is the engine performance for a given handle setting. In practice apart from the engine mass flow, in order to determine the overall engine performance, total thermodynamic quantities (such as total pressure and total temperature) and exit velocity are sufficient.

The intake performance under the assumption that the flow in not subjected to any kind of heat exchange with the environment, is fully determined as soon as the pressure recovery factor is determined. The pressure recovery factor can be read from an intake performance map where it is plotted against various intake inlet Mach numbers. Alternatively it can be calculated through an empirical correlation or it can be input directly as a value emerging purely from experience.

For the compressor, the performance map is usually plotted as the pressure ratio against the non dimensional air flow along lines of constant non-dimensional or relative rotational speed. Constant isentropic or adiabatic efficiency contours appear on the same graph.

Turbine's performance map is usually plotted as the expansion ratio or non-dimensional enthalpy drop versus the non-dimensional mass flow for constant non-dimensional rotational speed lines. The efficiency is usually plotted on a separate graph against the same parameters (non-dimensional mass flow and non dimensional rotational speed). A typical nozzle performance map includes the flow and discharge pressure coefficients as a function of the pressure ratio.

Finally concerning the combustion chamber, many kind of maps may be used, for the determination of the exit total temperature, the fuel flow (one of the previous two parameters is usually given as the engine handle) and the fuel flow composition. Similarly, total pressure loss can also be determined through performance maps or alternatively by the use of an empirical correlation based on hot and cold pressure loss coefficients.

The aim of the 0-D analysis is to fully determine pressures and temperatures as well as the other engine performance parameters throughout the engine, under any given set of boundary conditions. The solution is carried out at a certain off-design point which is specified by a user defined value of an appropriate engine handle. Engine handles can be the fuel flow, the rotational speed, or the turbine inlet temperature. The solution scheme, making use of the conservation of mass, energy and power, is searching for the set of operating points on each component's performance map with the aim of achieving the mechanical and thermodynam-ical engine matching.

At any given flight condition the Mach number and the altitude will determine the airflow at the inlet plane of the first compressor in the row. Subsequently, based upon the compressor's rotational speed and pressure ratio which are fixed once the operating point is fixed on the performance characteristic map, the compressor may not be capable of passing the amount of airflow coming from the intake duct. This gives rise to a flow imbalance. The same holds for all engine's compressors installed downstream. Similarly for the engine combustion chamber, given the burner exit temperature or fuel flow the exit pressure, the airflow at the first turbine face can be determined. The turbine may or may not be able to pass this airflow, depending on its operating point. In a similar way, the same flow imbalances may rise to all turbines and nozzles downstream. Apart from the flow imbalances, mechanical power imbalances may also exist, between a turbine's generated power and the coupled compressor absorbed power, for instance.

All these imbalances are treated as errors. The number of these errors varies according to the engine type. In any case a certain number of equations emerging from the conservation laws are needed for the engine to be balanced mechanically and thermodynamically. Various 0-D solution schemes can be employed for the error minimization. The error minimization process is realized by shifting the operating points on the components performance maps. There are three types of iterative solution schemes that can be used:

• The concentric iterative solution scheme.

• The crossover iterative solution scheme.

• The simultaneous iterative solution scheme.

According to the first two solution schemes the error appearing after the application of a certain conservation law, is minimized by changing a certain engine parameter. According to the third solution scheme, all errors are minimized simultaneously by changing simultaneously a pre-specified set of engine parameters. From the mathematical point of view the first two methods are much simpler. A simple trial and error iterative solution scheme for each air flow and power imbalance is enough. For the third method a more complicated mathematical solution scheme is required, like for instance a multi-dimensional Newton-Raphson method. The simultaneous iteration however is more stable than the previous two but higher computational power is required.

The current simulation program is based on the first type of iterative method, the concentric iteration. The operation of a single spool turbojet engine is simulated under any given set of boundary conditions. The engine rotational speed is set to be the engine handle, the parameter that is employed by the user to set the off-design condition.

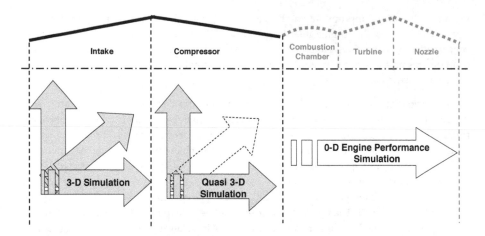

Figure 7. Intake – Gas Turbine Variable Dimension Performance Simulation Method.

5. Gas turbine engine simulation

5.1. Intake — Gas turbine variable dimension performance simulation method

As it was mentioned in the introductory chapter a single calculation method whichever that could be, cannot cover the need for a "light", in terms of computational load, gas turbine engine performance simulation code, able of addressing engine operation under non-uniform inlet conditions. A synthesis of several computational methods is therefore inevitable. The structural elements of the synthesized calculation method have already been described in the preceding corresponding sections.

Given the fact that flow within the intake and mostly within the compressor is redistributed any type of pressure or temperature distortion at the intake inlet practically vanishes by the time the working medium enters the combustion chamber. Even if any pressure or temperature gradients are persisting till that engine station, given the intense mixing that takes place within the combustion chamber, gradients of thermodynamical properties are redefined on a completely new basis, the combustion process. Consequently a 0-D analysis method downstream of the compressor is sufficient. Generally speaking the use of 2-D or 3-D flow simulation methods and their derived tools at hot section of the engine could only be justified in the case of detailed design.

On the contrary, the performance of the cold engine section, namely the intake and the compressor, is affected by the uneven flow properties distribution at the intake inlet. Regarding the intake section, the flow remains three dimensional and a 3-D flow simulation tool should be used in order to convey the flow information down to the compressor. Despite the fact flow non-uniformities are to a certain extent attenuated until they reach the compressor section, applying a 0-D or an 1-D flow simulation method will discard most of the information and will lead to highly inaccurate results.

5.2. Proposed variable dimension simulation

The current simulation technique was built with the aim of embracing the combined – coupled performance of the intake and the compressor. In reality the compressor map that comes as an intermediate output of the intake – compressor simulation, represents not only the compressor performance over its operating range as it is usually seen, but it also includes indirectly the intake performance. The simulation focuses on a certain compressor operating point defined by a certain rotational speed value and a mass flow value falling in the range of the selected rotational speed. Once the intake simulation has converged under a certain inlet boundary condition set, the flow-field on the compressor inlet plane is practically defined. 3-D flow calculation methods for compressor flow simulation have all been excluded mainly due to their high computational resources demand. In other words the derived compressor inlet boundary conditions have to be decomposed in order to reduce their dimensions. Two components are considered:

i.   Variation along the radial direction

ii.   Variation along the circumferential direction.

The derived engine flow simulation tool was designed with the intention of assessing the impact of 3-D pressure profiles on the engine performance. Therefore the distinction was made on the basis of the effect that each component has on compressor performance. The main effect of circumferential total pressure distortion on compressor performance is the surge line shift. On the contrary radial pressure distortion has the exact opposite effect, especially for multistage compressors. According to a NASA report [30] radial distortion did not appear to change the flow in the stages that control stall, because of the rapid attenuation of the distortion within the compressor. In other words surge line position is not severely affected by this type of distortion. Based on the above observations, the SLC code under the radial profile of boundary conditions, computes the pressure ratio and the isentropic efficiency corresponding to the initially specified mass flow and rotational speed. While the same rotational speed is retained, a small (positive of negative) increment is added to the mass flow and the whole process starting from the intake simulation, is repeated again until the entire mass flow range for this rotational speed is covered. Subsequently a small increment (positive or negative) is added to the rotational speed and the whole process is repeated again until a certain range of non – dimensional speed lines is covered. By the end of the above described process the full compressor – intake performance characteristic map will have been obtained. The compressor map as extracted from the SLC code is fed as an input to the "extended compressor in parallel" code. The number of circumferential sectors, their extent, as well as their average pressure, should be specified as an input. The exact number of sectors selected in order for the circumferentially distorted pattern be described in best, should ideally be defined through a sensitivity analysis. Given the above input, the surge line shifts will be predicted by the "extended compressor in parallel" code. At this point the performance of the intake – compressor is fully determined. The resulted compressor performance map, together with the input data referring to the remaining engine components, is fed into the 0-D simulation code for the engine performance to be defined.

On figure 8 it is demonstrated on the compressor map, a snapshot of the response of a small single spool turbojet engine operating behind a generic intake under certain sets of non-uniform inlet conditions.

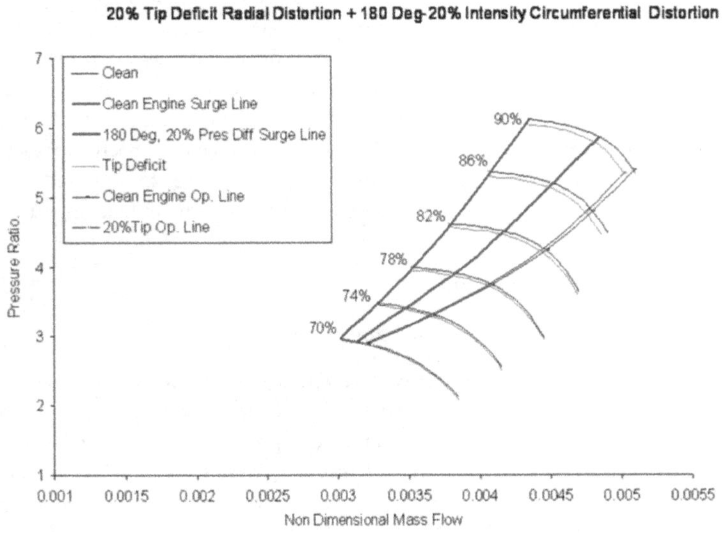

Figure 8. Compressor map shifts under various inlet conditions.

6. Closure

Gas turbine engine performance simulation is a very wide research sector of gas turbine related technology. The current chapter focused on the intermediate fidelity – relative low computational power consumption type of methods and derived tools. These tools in order to meet the contradictive requirements for 3-D flow treatment and fast convergence are based on more than one method. Moreover they make use of stored performance related information that come out from experiment and/or high fidelity flow simulation. Such simulation tools prove very convenient to gas turbine engine manufacturers as they can "multiply" the value of their numerous stored experimental data. It is the author's view that such hybrid simulation methods will on one hand be gaining increasing attention by the manufacturers, while on the other hand they consist a brilliant research field for creative thinking a combination which will lead to very valuable applications in the future.

Nomenclature

Abbreviations

CFD - Computational Fluid Dynamics

RANS - Reynolds Averaged Navier Stokes

REE - Radial Equilibrium Equation

SLC - Streamline Curvature

VLM - Vortex Lattice Method

0-D - Zero-Dimensional

1-D - One-Dimensional

2-D - Two-Dimensional

3-D - Three-Dimensional

Symbols

F - Force

H - Enthalpy

I - Rothalpy

P - Pressure

S - Entropy

T - Temperature

V - Absolute air velocity

W - Relative velocity

c - Constant of integration

i,j,k - Unit vectors

m - Meridional direction

n - Normal (to the meridional) direction

r - Radius, radial direction

rc - Radius of curvature

s - Tangential along the blade edge, direction

z - Axial direction

Greek Symbols

Γ - Circulation

α - Absolute flow angle

β - Relative flow angle

γ - Sweep angle

ε - streamline slope angle

λ - Lean angle

ϱ - Density

ω - Angular speed, Loss factor

Subscripts

D - Drag

P - Pressure

Re - Reynolds

j - Streamline counter

m - Meridional direction

n - Normal direction

prof - Profile

r - Radial direction

sec - Secondary

sh - Shock

tot - Total

w - Whirl direction

z - Axial direction

Author details

Ioannis Templalexis*

Hellenic Air Force Academy Department of Aeronautical Sciences Section of Thermodynamics, Propulsion and Power Systems , Greece

References

[1] www.gasturb.de

[2] M.G. Turner, J.A. Reed, R. Ryder and J.P. Veres, "Multi-Fidelity Simulation of a Turbofan Engine with Results Zoomed into Mini-Maps for a Zero-D Cycle Simulation," ASME Paper No. GT2004-53956, Vienna, Austria, June, 2004 (Also NASA TM-2004-213076)

[3] Alexiou A., Baalbergen E., Mathioudakis K., Kogenhop O., Arendsen P., "Advanced Capabilities for Gas Turbine Engine Performance Simulations", ASME paper GT2007-27086.

[4] Alexiou A., Mathioudakis K., "Gas Turbine Engine Performance Model Applications Using An Object-Oriented Simulation Tool", ASME paper GT2006-90339.

[5] Alan Hale, Milt Davis, and Jim Sirbaugh, "A Numerical Simulation Capability for Analysis of Aircraft Inlet-Engine Compatibility" J. Eng. Gas Turbines Power 128, 473 (2006)

[6] Pachidis, V., Pilidis, P., Talhouarn, F., Kalfas, A. and Templalexis, I., "A fully integrated approach to component zooming using computational fluid dynamics", Transactions of the ASME, Journal of Engineering for Gas Turbines and Power, Vol. 128, No.3, p. 579, July 2006.

[7] Pachidis, V., Pilidis, P., Marinai, L., Templalexis, I., "Towards a full two dimensional gas turbine performance simulation", Proceedings of the RASoc, The Aeronautical Journal, AJ-3127, June 2007.

[8] Templalexis, I., Pilidis, P., Pachidis, V. and Kotsiopoulos, P., "Development of a 2D compressor streamline curvature code", Transactions of the ASME, Journal of Turbomachinery, TURBO-06-1178, Vol. 129, Issue 4, October 2007.

[9] [136] Veres J.P. Overview Of High Fidelity Modelling Activities in The Numerical Propulsion System Simulations Project, NASA TM 2002-211351.

[10] Panel Methods an Introduction, NASA Technical Paper 2995, Decmber 1990.

[11] Katz, J. Plotkin A. Low Speed Aerodynamics From Wing Theory To Panel Methods. Mc Graw Hill Series in Aeronautical and Aerospace Engineering, 1989.

[12] Wu, C.H., 1952, "A General Theory of Three Dimensional Flow in Subsonic and Supersonic Turbomachines of Axial, Radial and Mixed Flow", NACA TN-2604.

[13] Smith, L.H. 1966, "The Radial Equilibrium Equation of Turbomachinery", Trans. A.S.M.E., Series A, Vol 88.

[14] Novak, R.A. "Streamline Curvature Computing Procedures for Fluid Flow Problems.", A.S.M.E. paper 66-WA/GT-3.

[15] Silvester, M.E. and Hetherington, R., 1966, "Three Dimensional Compressible Flow Through Axial Flow Turbomachines." Numerical Analysis an Introduction, Academic Press.

[16] Frost, D.H., 1972, "A Streamline Curvature Through-Flow Computer Program for Analysing the Flow Through Axial-Flow Turbomachines.", Aeronautical Research Council, R&M 3687.

[17] Senoo, Y. and Nakase, Y., 1972 "An Analysis of Flow Through a Mixed Flow Impeller.", ASME J. of Eng. for Power, pp. 43-50.

[18] Novak, R.A. and Hearsey, R.M., 1977, "A Nearly Three-Dimensional Intrablade Computing System for Turbomachinery.", J. of Fluids Eng. pp. 154-166.

[19] I.K. Jennions, P. Stow, 1985, "A Quasi-Three-Dimentional Turbomachinery Blade Desigh System: Part I – Throughflow Analysis.", J. of Eng. for Gas Turbines and Power, Vol. 107, pp. 301-307.

[20] Shaalan, M. R. A. and Daneshyar, H., 1972, "A Critical Assessment of Methods of Calculating Slope and Curvature of Streamlines in Fluid Flow Problems." Proc. of the Institution of Mechanical Engineers, Vol. 186.

[21] Wilkinson, D. H., 1969-70, "Stability, Convergence and Accuracy of 2-D Streamline Curvature Methods Using Quasi-Orthogonals." Proc. of the Institution of Mechanical Engineers, Vol. 184.

[22] M.V. Casey, 1984, "A Streamline Curvature Throughflow Method for Radial Turbocompressors.", C57/84 IMechE.

[23] Denton, J. D., 1978, "ThroughFlow Calculations for Transonic Axial Flow Turbines.", J. of Eng. for Power, Vol.100, pp. 212-218.

[24] Horlock, J. H., 1971, "On entropy production in adiabatic flow in Turbomachines", ASME Paper No 71-F3-3.

[25] Horlock, J. H. and March, H., "Flow models for turbomachines.", Journal of Mechanical Engineering.

[26] Arthur, J. Wennerstorm, 2000, "Design of Highly Loaded Axial Flow Fans and Compressors", Consepts ETI Press, pp93-99.

[27] Templalexis, I., Pachidis, V., Pilidis, P., and Kotsiopoulos, P., "The Effect of blade Lean on the Solution of the Full Radial Equilibrium Equation", GT2008-50259, ASME Turbo Expo, Power For Land, Sea and Air, Berlin, Germany, June 2008.

[28] Miller G. R., Lewis, G. W. and Hartmann M. J., 1961 "Shock Losses In Transonic Rotor Rows", Transactions of the ASME, Journal of Engineering for Power, Vol. 83, pages 235-242.

[29] Creveling H. F. and Carmody R. H., 1968 "Axial Flow Compressor Computer Program for Calculating Off-Design Performance (Program IV)", General Motors, Allison Division, Indianapolis, Prepared for NASA, Report CR-72427.

[30] Hager, R., D. 1977, Analysis Of Internal Flow Of J85-13 Multistage Compressor. NASA TM X-3513, Washington June 3, 1955.

[31] Calogeras, J., E., Johnsen, R., L., Burstadt, P., L., "Effect Of Screen Induced Total Pressure Distortion On Axial Flow Compressor Stability." NASA TM X-3017. Lewis research center, National aeronautics and space administration, May 1974.

[32] Milner, E., J., 1977, Performance And Stability Of A J85-13 Compressor With Distorted Inlet Flow., NASA TM X-3515. Lewis research center, National aeronautics and space administration, May 1977.

[33] Pearson H., McKenzie, A., B., Wakes In Axial Compressors. J. R. Aeronautical Society, Vol 63, No 583 July 1959, pp.415-416.

[34] Mazzawy, R., S., Multiple Segment Parallel Compressor Model For Circumferential Flow Distortion. ASME Journal of engineering for power, Vol 99, Apr.1977, pp228-246.

[35] Steenken, W., G., Modeling Compressor Component Stability Characteristics-Effects Of Inlet Distortion And Fan Bypass Duct Disturbances. Engine handling, AGARD CP-324, Oct. 1982.

[36] Baghdadi, S. and Lucke, J., E., Compressor Stability Analysis, ASME Paper No. 81-WA/FE-18.

[37] Hynes, T., P., Greitzer, E., M., A Method For Assessing Effects Of Circumferential Flow Distortion On Compressor Stability, Journal of turbomachinery, Vol. 109, July 1987, pp.371-379.

[38] Plourde, G., A., Stenning A., H., The Attenuation Of Circumferential Inlet Distortion In Multi-Stage Axial Compressors. AIAA 3rd Propulsion joint specialist conference, Washington, D., C., July 17-21, 1967.

[39] Longley, J. P., Greitzer, E., M., Inlet Distortion Effects In Aircraft Propulsion System Integration. Trans of ASME ASME/94-GT-220, 1994.

Chapter 2

Micro Gas Turbine Engine: A Review

Marco Antônio Rosa do Nascimento,
Lucilene de Oliveira Rodrigues,
Eraldo Cruz dos Santos, Eli Eber Batista Gomes,
Fagner Luis Goulart Dias,
Elkin Iván Gutiérrez Velásques and
Rubén Alexis Miranda Carrillo

Additional information is available at the end of the chapter

1. Introduction

Microturbines are energy generators whose capacity ranges from 15 to 300 kW. Their basic principle comes from open cycle gas turbines, although they present several typical features, such as: variable speed, high speed operation, compact size, simple operability, easy installation, low maintenance, air bearings, low NO_x emissions and usually a recuperator (Hamilton, 2001).

Microturbines came into the automotive market between 1950 and 1970. The first microturbines were based on gas turbine designed to be used in generators of missile launching stations, aircraft and bus engines, among other commercial means of transport. The use of this equipment in the energy market increased between 1980 and 1990, when the demand for distributed generating technologies increased as well (LISS, 1999).

Distributed generation systems may prove more attractive in a competitive market to those seeking to increase reliability and gain independence by self-generating. Manufacturers of gas and liquid-fueled microturbines and advanced turbine systems have bench test results showing that they will either meet or beat current emission goals for nitrogen oxides (NOX) and other pollutants (Hamilton, 2001). Air quality regulation agencies need to account for this technological innovation. Emission control technologies and regulations for distributed generation system are not yet precisely defined. However, control technologies that could reduce emissions from fossil-fueled components of a distributed generation system to levels similar to other traditional fossil-fueled generation equipment are already available.

Combustion processes can result in the formation of significant amounts of nitrogen dioxide (NO_2) and carbon monoxide (CO). Some manufacturers of microturbines have developed advanced combustion technologies to minimize the formation of these pollutants. They have assured low emissions levels from microturbines fueled with gaseous and liquid fuels.

2. History

In fact, the technology of microturbines is not new, as researches on this subject can be found since 1970, when the automotive industry viewed the possibility of using microturbines to replace traditional reciprocating piston engines. However, for a variety of reasons, microturbines did not achieve great success in the automotive segment. The first generation of microturbines was based on turbines originally designed for commercial applications in generating electricity for airplanes, buses, and other means of commercial transportation.

The interest in the market for stationary power spread in the mid-1980 and accelerated in the 1990s, with its reuse in the automobile market in hybrid vehicles and when demand for distributed generation increased (Liss, 1999). Currently, the operation of hybrid vehicles through a microturbine connected to an electric motor, have received special attention from some of the major car manufacturers such as Ford, and research centers (Barker, 1997).

In 1978, Allison began a project aimed at the development and construction of generating groups for military applications, driven by small gas turbines. The main results obtained during testing of these generators revealed: reduction in fuel consumption of 180 l/h to 60 l/h, compared with previous models, frequency stability of about 1%, noise levels below 90 dB and the possibility of using different fuels (diesel, gasoline, etc.). In 1981, a batch with 200 generators was delivered to the U.S. Army, and since then, more than 2,000 units have been provided to integrate the system of electricity generation for Patriot missile launchers (Patriot Systems) (Scott, 2000).

The deregulation of the electricity market in the United States began in 1978 when the Power Utility Regulatory Policy Act (PURPA) revolutionized the energy market in the United States, breaking the monopoly of the electricity generation sector, enabling the beginning of the expansion of distributed generation. Since then there has been a significant increase in the proportion of independent generation in the country and, according to a projection made in 1999 by the Gas Research Institute (GRI), this in-house production should reach 35% in 2015 (Gri, 1999).

With a new market structure, i.e., with the possibility of attracting small consumers of energy, microturbines began to be the target of intense research. Already in 1980, under the support of the Gas Research Institute, a program entitled Advanced Energy System (AES) was initiated with a view to develop a small gas turbine, with typical features of aviation turbine, rated at 50 kW and equipped with a heat recovery for a system cogeneration. The program was abandoned around 1990 by the Gas Research Institute, on the grounds of problems with the final cost of the product (Watts, 1999). Since then, the Gas Research Institute began to support new projects in partnership with several companies, such as the Northern Research & Engineering Energy Systems, also supporting the first efforts of Capstone Turbine Corporation (still under the name of its precursor, NoMac Energy Systems) (Gri, 1999).

Some companies in the United States, England and Sweden have recently introduced in the world market commercial units of microturbines. Among these companies are: AlliedSignal, Elliott Energy Systems, Capstone, Ingersoll-Rand Energy Systems & Power Recuperators WorksTM, Turbec, Browman Power and ABB Distributed Generation & Volvo Aero Corporation.

3. State-of-the-art microturbines

AlliedSignal microturbine has shaft configuration, works with cycle Regenerative open Brayton, its bearings are pneumatic and it has a drive direct current - alternating current (DC/AC) 50/60 Hz (the frequency is reduced from about 1,200 to 50 Hz or 60 Hz) and the compressor and turbine are the radial single stage. The heat transfer efficiency of this stainless steel regenerator is 80-90%. Besides working with diesel oil and natural gas, this microturbine can burn naphtha, methane, propane, gasoline, and synthetic gas. Its noise level is estimated at 65 dB. A commercial prototype of 75 kW was designed for a 30% efficiency and its installed cost is estimated from $ 22,500 to 30,000 (Biasi, 1998).

Elliott Energy Systems (a subsidiary of Elliott Turbomachinery Company) has a manufacturing and assembly unit in Stuart, Florida with a production capacity of 4,000 units per year. According to Richard Sanders, executive vice president of sales and marketing, Elliott has launched two commercial prototypes: a 45 kW microturbine (TA-45model) and another 80 kW (TA-80), and later, a 200 kW microturbine (TA-200). The TA-45 model is rated at 45 kW (Figure 1) at ISO conditions and its main difference from other manufacturers is that it has oil lubricated bearings and a system starting at 24 volts, which, according to Sanders, is unique to microturbines. The TA-80 and TA-200 microturbines models are similar to the TA-45 model. All three can generate electricity in 120/208/240V and can work with different fuels: natural gas, diesel, kerosene, alcohol, gasoline, propane, methanol and ethanol (Biasi, 1998).

The development works of the components has taken the Capstone in the 90's, build and tested a prototype of a 24 kW microturbine in 1994. And in 1996, Capstone made a project consisting of 37 prototypes for field testing. According to Biasi, 1998, Paul Craig, the President of Capstone Turbine Corporation, expected the 30-kW business model to have a cost of about $ 500/kW (installed microturbine) and a generation cost of $ 45-50/MWh. Figure 2 shows Capstone microturbine, model C65, which is already commercially available.

Figure 1. Elliott Energy Systems Microturbine, TA-45 model.

Figure 2. Capstone microturbine, model C65 (Capstone, 2012).

Four Honeywell Power Systems microturbines of 70 kW each were, until 2001, being tested in the Jamacha Landfill in New Hampshire - United States. The gas produced in the landfills was about 37% methane, carbon dioxide and air. The gas was cooled to about 14 °C to remove moisture and impurities and then compressed to about 550 kPa for the microturbine power. For the first 3 minutes of turbine operation, the fuel feed was carried out with propane. The system operated in parallel and exported electricity to San Diego Gas & Electric. In September 2001, Honeywell decided to stop manufacturing microturbines and uninstalled the four microturbines from the Jamacha Landfill, Figure 3. Until that time, the microturbines operated for 2000 hours, without showing degradation in performance. Then, the microturbines from Honeywell Microturbines were replaced by turbines with the same capacity from Ingersoll-Rand Power Works™, as shown in Figure 4 (Pierce, 2002).

In order to develop a new generation of microturbines, in 1998 ABB Distributed Generation established a 50/50 joint venture with Volvo Aero Corporation. This partnership joined the experience of Volvo gas turbine for hybrid electric vehicles with the experience of ABB in the generation and energy conversion at high frequency. This joint venture resulted in the development of a microturbine for cogeneration. Operating on natural gas, the MT100 microturbine generates 100 kW of electricity and 152 kW of thermal energy (hot water). As other manufacturers of microturbines, the MT100 has a frequency converter that allows the generator to operate at variable speed.

Table 1. brings is a summary of the main features of microturbine leading manufacturers.

Figure 3. Ingersoll-Rand Power WorksTM installed on the Jamacha Landfill - United States.

Figure 4. Prototype Ingersoll-Rand Power Works™ installed on Jamacha Landfill - United States.

Model	Manufacturers	Power Output	Set	Total Efficiency (LHV)	Pressure Ratio	TET	Nominal Speed
		kW		%		°C	Rpm
-	AlliedSignal	75	A Shaft	30 (HHV)	3.8	871	85,000
TA 45	Elliott Energy System	45	A Shaft	30	-	871	-
TA 80	Elliott Energy System	80	A Shaft	30	-	871	68,000
TA 200	Elliott Energy System	200	A Shaft	30	-	871	43,000
C30	Capstone	30	A Shaft	28		871	96,000
C65	Capstone	65	A Shaft	29		871	85,000
C200 HP	Capstone	200	A Shaft	33		870	45,000
-	Power Works™	70	Two Shafts	30 (HHV)	3	704	-
MT 100	ABB	100	A Shaft	30	4.5	950	70,000

Table 1. Technical characteristics of leading microturbine manufacturers.

Microturbines are lower power machines with different applications than larger gas turbines, having typically the following characteristics:

- Variable rotation: the turbine variable speed is between 30,000 and 120,000 rpm depending on the manufacturer;

- High frequency electric alternator: the generator operates with a converter for AC/DC. In addition, the alternator itself is the engine starter;

- Reliability: some microturbines have already reached 25,000 hours of operation (approximately three years) including shutdown and maintenance;

- Simplicity: the generator is placed in the same turbine shaft being relatively easy to be manufactured and maintained. Moreover, it presents a great potential for inexpensive and large scale manufacturing;

- Compact: easy installation and maintenance;

- High noise levels: to reduce noise levels during operation, microturbines require a specific acoustic system;

- Air-cooled bearings: the use of air bearings avoid lubricants contamination by combustion products, prolongs the equipment useful life and reduces maintenance costs;

- Retrieve: microturbine manufacturers generally use heat recovery of exhaust gas to heat the air intake of the combustion chamber, thus achieving a thermal efficiency of 30%.

4. Configuration

Microturbines have similar set-up of small, medium and large size gas turbines, as described by Nascimento and Santos (2011), i.e., microturbines are formed by an assembly of a compressor, a combustion chamber and a turbine, as shown in the simplified scheme of Figure 5.

State-of-the-art microturbines have markedly improved in the last years. Several microturbines have been developed by manufacturers with different configurations. Their configuration depends on the application, although they usually consist of a single-shaft microturbine, annular combustor, single stage radial flow compressor and expander, and a recuperator or not. The optimum microturbine rotational speeds at typical power ratings are between 60 to 90,000 rpm and pressure ratio of 3 or 4 : 1, in a single stage.

Gas microturbines have the same basic operation principle as open cycle gas turbines (Brayton open cycle). Figure 5 shows the Brayton open cycle. In this cycle the air is compressed by the compressor, going through the combustion chamber where it receives energy from the fuel and thus raising its temperature. Leaving the combustion chamber, the high temperature working fluid is directed to the turbine, where it is expanded by supplying power to the compressor and for the electric generator or other equipment available.

Microturbines are a technology based cycle with or without recuperation. To produce an acceptable efficiency, the heat in the turbine exhaust system must be partially recovered and
used to preheat the turbine air supply before it enters the combustor, using an air-to-air heat
exchanger called recuperator or regenerator. This allows the net cycle efficiency to be increased to as much as 30% while the average net efficiency of unrecovered microturbines is
17 % (Rodgers et. al., 2001a).

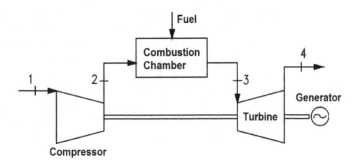

Figure 5. Gas turbine system scheme of a simple open cycle.

As well as in gas turbines, the maximum net power provided by a microturbine is limited
by the temperature the material of the turbine can support, associated with the cooling technology and service life required. The two main factors affecting the performance of microturbines are: components efficiency and gases temperature at the turbine inlet.

Furthermore, microturbines usually employ permanent magnet variable-speed alternators
generating very high frequency alternating current which must be first rectified and then
converted to AC to match the required supply frequency.

Capstone Microturbines, shown in Figure 6, uses a lean premix combustion system to achieve low emissions levels at a full power range. Lean premix operation requires operating at
high air-fuel ratio within the primary combustion zone. The large amount of air is thorough-
ly mixed with fuel before combustion. This premixing of air and fuel enables clean combustion to occur at a relatively low temperature. Injectors control the air-fuel ratio and the air-
fuel mixture in the primary zone to ensure that the optimal temperature is achieved for the
NO_x minimization. The higher air-fuel ratio results in a lower flame temperature, which
leads to lower NO_x levels. In order to achieve low levels of CO and Hydrocarbons simultaneously with low NO_x levels, the air-fuel mixture is retained in the combustion chamber for
a relatively long period. This process allows for a more complete combustion of CO and Hydrocarbons (Capstone, 2000).

In addition, the exhaust of microturbines can be used in direct heating or as an air pre-heat-
er for downstream burners, once it has a high concentration of oxygen. Clean burning combustion is the key to both low emissions and highly durable recuperator designs.

The most effective fuel to minimize emissions is clearly natural gas. Natural gas is also the fuel choice for small businesses. Usually the natural gas requires compression to the ambient pressure at the compressor inlet of the microturbine. The compressor outlet pressure is nominally three to four atmospheres.

Capstone microturbine control and power electronic systems allow for different operation modes, such as: grid connect, stand-alone, dual mode and multiple units for potentially enhanced reliability, operating with gas, liquid fuels and biogas. In grid connect, the system follows the voltage and the frequency from the grid. Grid connect applications include base load, peak shaving and load following. One of the key aspects of a grid connect system is that the synchronization and the protective relay functions required to reliably and safely interconnect with the grid can be integrated directly into the microturbine control and power electronic systems. This capability eliminates the need for very expensive and cumbersome external equipment needed in conventional generation technologies (Rodgers et. al., 2001a). In the stand-alone mode, the system behaves as an independent voltage source and supplies the current demanded by the load. Capstone microturbine when equipped with the stand-alone option includes a large battery used for unassisted black start of the turbine engine and for transient electrical load management.

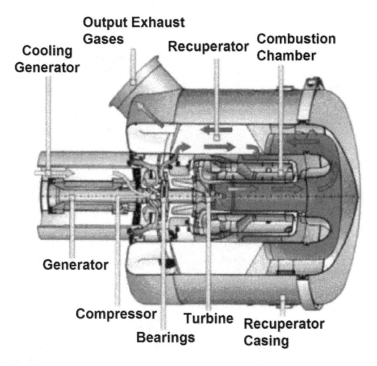

Figure 6. Parts of a Capstone microturbine.

In both operational mode, that is, the grid connect and the stand-alone, the microturbine can also be designed to automatically switch between these two modes. This type of functionality is extremely useful in a wide variety of applications, and is commonly referred to as dual mode operation. Besides, the microturbines can be configured to operate in parallel with other distributed generation systems in order to obtain a larger power generation system. This capability can be built directly into the system and does not require the use of any external synchronizing equipment.

Some microturbines can operate with different fuels. The flexibility and the adaptability enabled by digital control software allow this to happen with no significant changes to the hardware. Power generation systems create large amounts of heat in the process of converting fuel into electricity. For the average utility-size power plant, more than two-thirds of the energy content of the input fuel is converted into heat. Conventional power plants discard this waste heat, however, distributed generation technologies, due to their load-appropriate size and sitting, enable this heat to be recovered. Cogeneration systems can produce heat and electricity at or near the load side. Cogeneration plants usually have up to 85% of efficiency and operation cost lower than other applications. Small cogeneration systems usually use reciprocating engines although microturbines have showed to be a good option for this application. The hot exhaust gas from microturbines is available for cogeneration applications. Recovered heat can be used for hot water heating or low-pressure steam applications.

5. Experimental set-up for microturbine

To perform tests in microturbines, a test bench was built in the Laboratory of Gas Turbines and Gasification of the Institute of Mechanical Engineering, Federal University of Itajubá - IEM/UNIFEI. This bench was composed of a 30 kW regenerative cycle diesel single shaft gas microturbine engine with annular combustion chamber and radial turbomachineries, as shown in Figure 7, and was configured to operate with liquid fuel.

The microturbine engine was tested while in operation with automotive ethanol and pure diesel, respectively. Thermal and electrical parameters, such as mass flows, temperature, composition of exhaust gases and generated power were constantly measured during the tests.

Figure 8 shows the scheme of the microturbine with the measuring points. The microturbine engine was tested during operation with ethanol and diesel at steady state condition and at partial, medium and full loads.

As can be seen in Figure 8, all parameters assessed, during laboratory tests, were acquired and post-processed in a supervisory system developed in the laboratory UNIFEI.

In order to establish whether the fuels were able to feed the engine without presenting any problems regarding the fuel injection system, the kinematic viscosity of each fuel was measured. The composition of the emission gases and the thermal variables were also measured at medium and full loads for each fuel, and their results are presented below. All tests were performed in the grid connection mode.

Figure 7. Capstone microturbine in the laboratory at UNIFEI.

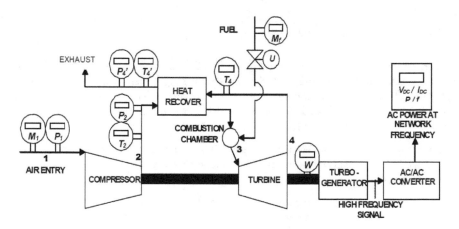

Figure 8. Schematic representation of the test rig and the data acquisition system.

This microturbine is mainly used for primer power generation or emergency and can work with a variety of liquid fuels. This microturbine uses a recovery cycle to improve its efficiency during operation, due to a relatively low pressure, what facilitates the use of a single shaft radial compression and expansion [Cohen, et. al., (1996), Capstone, (2001), Roger, et. al., (2001b), Bolszo (2009)]. Table 2 shows the engine design characteristics at ISO condition.

Fuel Pressure	350 kPa
Power Output	29 kW NET (± 1)
Thermal Efficiency	26% (± 2)
Fuel HHV	45,144 kJ/kg
Fuel Flow	12 l/h
Exhaust Temperature	260 °C
Inlet Air Flow	16 Nm³/min
Rotational Speed	96000 rpm
Pressure Ratio	4

Table 2. Engine Performance data at ISO Condition.

For tracking and measuring the tests parameters a type of supervisory software was used in the test bench (given by the turbine manufacturer) along with the data acquisition and the post processing obtained during the tests.

The composition of the exhaust gases was measured in real time using an Ecoline 6000 gas analyzer, reporting the concentration of O_2, CO_2 and hydrocarbons (HC) in volume percentage (%v/v) and NO, CO, NO_2 and SO_2 (ppm) (Sierra, 2008). The fuel high heating value (HHV) was determined by a C-2000 IKA WORKS calorimeter. The accuracy, range and resolution of each instrument used during the tests are shown in Table 3.

Instrument		Range	Resolution	Accuracy
Fuel Flow		0-100 (l/h)	1.0 (ml)	±1.0 (%) scale
Temperature		0-350 (°C)	0.31 (°C)	±0.8 (%) scale
Pressure		0-10 (bar)	0.01 (bar)	±1.0 (%) scale
Power		0-45 (kW)	0.05 (kW)	±0.5 (%) scale
Calorimeter		--	--	±0.5 (%)
Gas Analyzer	CO (ppm)	0 - 20000	1	± 10 < 300 ± 4 (%) rdg < 2000 ± 10 (%) rdg "/> 2000
	NOx (ppm)	0 - 4000	1	± 5 < 100 ± 4 (%) rdg < 3000

Table 3. Accuracy of the measuring instruments.

5.1. Adjustments to the microturbine

Due to impurities in the gas or fuel, for instance, in the synthesis or biofuel, a redesign of the gas turbine combustor was necessary. For each type of fuel, a different kind of optimization was needed, in relation to the fuel low heating value (LHV).

To compensate for the lower heating value (LHV) of fuel gases, the fuel injection system must provide a much higher fuel rate than when operating with high heating values. Due to the high rate of mass flow of gas with LHV, the passage of fuel has a much larger cross section than the section corresponding to natural gas. Fuel pipes, control valves and stop valves have larger diameters and shall be designed to include an additional fuel blend, which consists of the final mixture of the recovered gas with natural gas and steam. The pressure drops and the size of the air spiral entering the flame tube must be adjusted to optimize the combustion process. The system must have high safety standards, so the flanges and the gaskets of the combustor and its connections must be safely welded. The system for low LHV must include:

- Fuel line for a LHV;

- Natural gas line;

- Steam line to reduce NO_x;

- Line blending of fuel for LHV;

- Line of nitrogen to purge;

- Lines pilot;

- Compressor;

- Combustion Chamber.

For safety reasons, the loading of the gas turbine to the rated load is accomplished through the use of the fuel reserve. The procedure for replacing the fuel reserve to the main tank is done automatically.

5.2. Tests on gas turbine using liquid fuel

The performance of a gas turbine is related to the local conditions of the installation and the environment, where pressure and temperature conditions are of great importance.

Due to the diesel low solubility at low temperature, tests with ethanol were performed without premix, and without the use of additives, which increased the cost of fuel.

According to the measuring methodology to be adopted to test gas turbines operating on liquids fuels, the physical-chemical properties of ethanol and diesel are shown in Table 4.

Table 4 also shows the fuel requirements established by the manufacturer of the tested gas turbine along with ASTM D6751 standard specifications for the testing of thermal performance. Regarding emissions a standard ISO 11042-1:1996 was used (NWAFOR, 2004).

Properties	Ethanol	Diesel	Fuel Limits	ASTM D6751
Sulfur (% mass)	0	0.20	0.05 <	< 0.05
Kinematic Viscosity @40 °C (mm²/s)	1.08	1.54	1.9 – 4.1	1.9 – 6
Density @ 25 °C (g/cm³)	0.786	0.838	0.75 – 0.95	-
Flash Point (°C)	13	60	38 - 66	"/> 130
Water (% Volume)	0.05	0.05	0.05	0.05
LHV (kJ/kg)	23,985.00	42,179.27		

Table 4. Ethanol and diesel physical-chemical characteristics.

The experimental determination of the ethanol heating value, kinematic viscosity and density were carried out according to ISO 1928-1976 and ASTM D1989-91 standards (ASME, 1997).

The use of different fuels implies the need of mass flow rate adjustments, according to its LHV and density, as without these adjustments, once established a load, the supply system would feed a quantity of fuel depending on the characteristics of the standard fuel (diesel). If the LHV of the new fuel is lower than standard, the gas turbine power could not reach the required demand.

Initially, the engine operated with conventional diesel fuel for a period of 20 minutes to reach a steady state condition for a load of 10 kW. After 20 minutes, the mass flow rates were changed to the fuel corresponding values. At this stage the fuel started to be replaced in order to increase the content of ethanol, by closing the diesel inlet valve and opening the ethanol valve. In order to ensure that all existing diesel power on the engine internal circuitry would be consumed, the engine was left running for 10 minutes with the same load operation, that is, 10 kW.

In order to check if the fuels were able to supply the engine, without causing problems to the fuel injection system, the kinematic viscosity of each fuel was measured. The composition of gas emissions and thermal parameters were also measured in total and average load for each fuel. This whole procedure was performed for the engine operating with loads of 5, 10, 15, 20, 25 and 30 kW in a grid connection mode.

Afterwards the emissions were measured with a gas analyzer, and the load of 5 kW increased. Ten minutes were necessary until it reached steady state again. Exhaust emissions were measured from the exhaust gases and, as mentioned before, the thermal performance data were stored in a personal computer (PC) unit coupled with a PLC (Programmable Logic Controller) data acquisition system, which carried out the data reading at every second.

When tests with ethanol were over, the engine was left running, in order to accomplish the purging of the remaining fuel. After that the engine was once again operated with diesel for ten minutes, and then disconnected and stopped.

6. Performance evaluation

The performance showed in this study was obtained from experimental tests at the Gas Turbine Laboratory of the Federal University of Itajubá (GOMES, 2002). Both natural gas and
liquid fuel Capstone microturbines and their respective fuel supplying and electrical connection systems were installed and a property measurement was used to obtain the behavior of microturbines operating at partial and full load.

6.1. Natural gas

The microturbine tested on natural gas was a Capstone 330 High Pressure. Table 5 gives the
technical information of this machine and the features of the natural gas used in the tests.
The natural gas microturbine was tested on the stand-alone mode supplying a resistive load.
These microturbines can record operational parameters (temperatures, pressures, fuel usage,
turbine speed, internal voltages/currents, status, and many others). Such data can be accessed with a computer or modem connected to an RS-232 port on the microturbine. To supplement these data, additional instrumentation was installed for the tests.

CAPSTONE Microturbine Features		
Model	330 (High Pressure)	
Full-Load Power (ISO Conditions)	30 kW	
Fuel	Natural Gas	
Fuel Pressure	358 – 379 kPa	
Fuel Flow*	12 m³/h	
Efficiency (LHV)*	27%	
Proprieties of Natural Gas (20 °C and 1 atm)		
Specific Mass	0.6165	
Low Heat Value	36,145	kJ/m³
High Heat Value	40,025	kJ/m³
Ambient Conditions		
Elevation	800	meters
Average Temperature	30	°C

Table 5. General conditions of the analysis

A large battery started the microturbine when disconnected from the grid, preventing any
sudden load increase or decrease in the electrical buffer during the stand-alone operation
(Capstone, 2001). The start-up took about 2 minutes and the speed was increased from 0
(zero) to 45,000 rpm, occasion when the microturbine started generating electricity. The ro-

tating components of the microturbine were mounted on a single shaft supported by air bearings and a spin at up 96,000 rpm. Figure 9 shows the speed behavior with the microturbine power output.

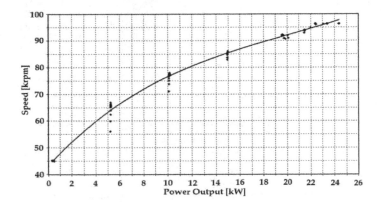

Figure 9. Microturbine speed at partial loads.

Capstone microturbine includes a recuperator which allows the microturbine efficiency to be improved. Figure 10 and 11 show respectively, the exhaust temperature and the efficiency behavior at partial loads. 27 % efficiency is possible at full load.

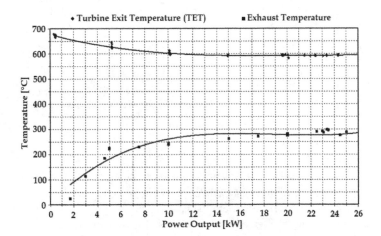

Figure 10. Microturbines exhaust temperature at partial loads.

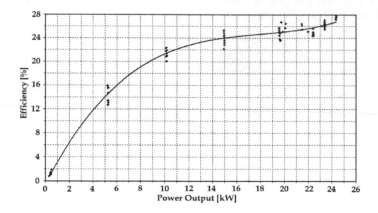

Figure 11. Microturbine efficiency at partial loads.

Figure 12 shows CO and NO$_x$ emissions behavior of a Capstone natural gas microturbine. Combustion occurs in three different steps. The first step is from start-up to about 5 kW. At this step CO formation decreases and emissions of NO$_x$ increase quickly.

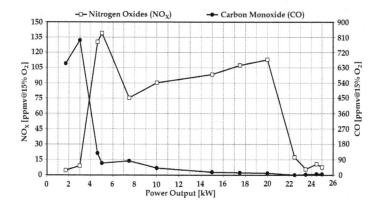

Figure 12. CO and NO$_x$ emissions of a natural gas microturbine at partial loads.

The second step is between 5 and 20 kW, as shown in Figure 12. In the second step the CO formation decreases continuously while emissions of NO$_x$ decrease at first, though increasing but it returns to increase softly slightly up to 113 ppmv. The last step begins at this point. At this step the lean-premix combustion occurs and the NO$_x$ formation diminishes to 5 ppmv.

Emissions of CO$_2$ depend on the fuel type and the system efficiency. Figure 13 shows CO$_2$ emissions of a Capstone natural gas microturbine.

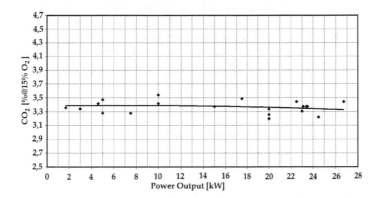

Figure 13. CO_2 emissions of a natural gas microturbine at partial loads.

6.2. Liquid fuel

The microturbine tested on diesel was a Capstone 330 Liquid Fuel. Table 6 gives the technical information of this machine and the features of the diesel used in the tests.

CAPSTONE Microturbine Features		
Model	330 (Liquid Fuel)	
Full-Load Power*	29 kW	
Fuel	Diesel #2 (ASTM D975)	
Fuel Pressure	35 – 70 kPa	
Fuel Flow*	12.5 l/h	
Efficiency (LHV)*	26%	
Proprieties of Liquid Fuel (20 °C and 1 atm)		
Specific Mass	0.848	
Low Heat Value	42,923	kJ/kg
High Heat Value	45,810	kJ/kg
Ambient Conditions		
Elevation	800	meters
Average Temperature	30	°C
* ISO Conditions		

Table 6. General conditions of the analysis

The liquid fuel microturbine was tested on the grid connect mode. These data can be accessed with a computer or modem connected to an RS-232 port on the microturbine. To supplement these data, additional instrumentation was installed for the tests. Figure 14 shows the turbine exit temperature and the exhaust temperature at partial loads. These temperatures are before and after the recuperator were used and their difference ranges from 300 to 450 °C.

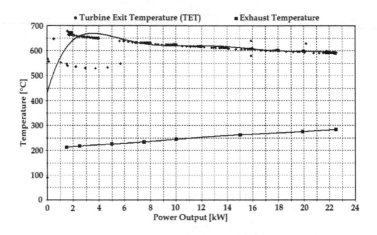

Figure 14. Microturbine exit and exhaust temperature at partial loads.

Figure 15 shows the liquid fuel microturbine efficiency at partial loads. Up to 24.5 % efficiency is possible at full load while the microturbine efficiency is at its highest when Capstone microturbines operate over an output range between 12 kW and full load.

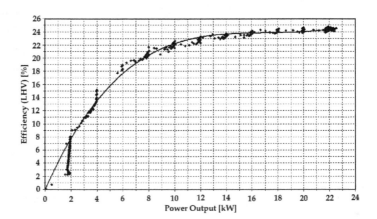

Figure 15. Microturbine efficiency at partial loads

Figure 16 shows the CO and NO$_X$ emissions behavior of a Capstone liquid fuel microturbine. The CO formation decreases, whereas emissions of NO$_X$ increase as the power output increases due to a rise in the flame temperature.

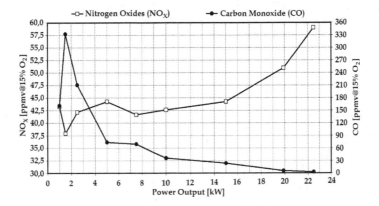

Figure 16. CO and NOX emissions from liquid fuel microturbine at partial loads.

Figure 17 shows the CO$_2$ and SO$_2$ emissions of a Capstone liquid fuel microturbine. The emissions depend considerably on the liquid fuel features. While SO$_2$ emissions are an important emission category for traditional electric utility companies, they are expected to be negligible for distributed generation technologies.

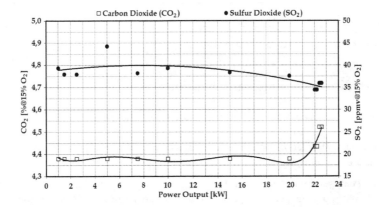

Figure 17. CO$_2$ and SO$_2$ emissions from liquid fuel microturbine at partial loads.

7. Microturbines and Internal combustion engine´s emissions

Table 7 and 8 compare emissions data from internal combustion engines and microturbines. In the absence of a post combustion device, such as a catalytic converter, reciprocating engines can have very high emission levels. Emission levels of microturbines are lower than levels of internal combustion engines as microturbines combustion is a continuous process which allows for a complete burning.

		ICE Natural Gas Without Control	ICE Natural Gas SCR	ICE Diesel Without Control	ICE Diesel SCR
Efficiency	% (HHV)	36%	29%	38%	38%
Nominal Power	kW	1,000	1,000	1,000	1,000
NO_x (@15%O_2)	g/MWh	998	227	9,888	2,132

SCR: Selective Catalytic Reduction.

Table 7. NO_x emissions of internal combustion engines (ICE) (Weston, *et. al.*, 2001)

FUEL		Natural Gas	Diesel
Efficiency*	% (LHV)	27	26
Nominal Power*	kW	30	29
CO (@15%O_2)**	g/MWh	210	80
NO_x (@15%O_2)**	g/MWh	520	280

* ISO Conditions; ** On Site Conditions (See Table 1)

Table 8. CO and NO_x emissions of Capstone microturbines

8. Case studies under Brazilian conditions

Due to the Brazilian governmental incentive to develop the gas industry, the feasibility of many natural gas applications has been doubted. Consequently, the demand for efficiently and environmentally friendly power generation technologies has increased. Many electricity consumers are considering producing their own electricity (Gomes, 2002).

This study analyses the possibility of natural gas application with Capstone microturbines in three cases of power generation: peak shaving in a small industry, base load in a gas station and a cogeneration system supplying buildings in a residential segment

Nowadays it is a trend on microturbines market to reduce investments. This paper analyses the influence of the investment cost of microturbines on the feasibility and cost of the generated electricity, being the cost of fuel a significant part of the electricity final price. The feasibility and the cost of the electricity generated with fuel were also assessed. This study used electric energy and natural gas prices charged by several electric power utility companies and gas distributors in Brazil at the time this study was being carried out (November, 2002). Table 9 shows the general conditions used in the cases studies.

Currency rate	2.6	R$/US$
Interest rate	10	% per year
CAPSTONE Microturbine Features		
Model	330 (High Pressure)	
Fuel	Natural Gas	
Proprieties of Natural Gas (20 °C and 1 atm)		
Specific Mass	0.602	
High Heat Value	39,304	kJ/m³

Table 9. General conditions of the analysis

8.1. Peak shaving case

Many consumers try to reduce their electricity consumption at peak hours due to its high price. If they can produce their electricity, they will reduce the amount of electricity purchased from utility companies at peak hours, without having to reduce their electricity consumption. Besides, power generation systems can improve the quality and reliability of the energy supplied by utility companies.

A study was carried out in four Brazilian regions, classified according to the price of natural gas charged by gas distributors of these regions, as shown in Table 10. Table 11 and Figure 18 show the conditions studied and the electricity demand supplied by utility companies with and without peak shaving.

	Brazilian States
1st Region	São Paulo (SP) and Rio de Janeiro (RJ)
2nd Region	Ceará (CE), Pernambuco (PE) and Paraíba (PB)
3rd Region	Rio Grande do Norte (RN)
4th Region	Others

Table 10. Brazilian regions analyzed in the peak shaving case

Commercial microturbines available in the Brazilian market are imported from the USA and investments feasibility depends on the currency rate, as can be seen in Table 9.

Model of microturbine	Capstone 330	
Number of microturbines	1	
Life time of microturbines	20	years
Net power (peak load)	28	kW
Microturbine installed cost	1.538	US$/kW
Natural gas consumption (HHV)	650	m³/month
Average price of natural gas (taxes included)	0.33-1.32	R$/m³

Table 11. Conditions of the peak shaving case

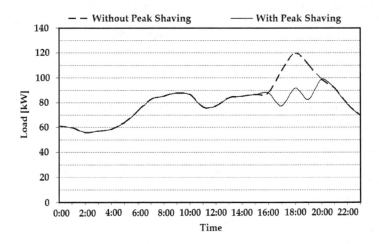

Figure 18. Electricity demand supplied by utility companies.

Table 12 displays the economical analysis of the peak shaving case. The investment is not feasible yet, as the payback period is very long. Rio Grande do Norte is the state where this business would be most interesting as payback is 8 years.

		SP - RJ	CE-PE-PB	RN	Other States
Total Investment *	US$	46,827	46,827	46,827	46,827
Annual cost**	US$/year	53,827	44,906	55,718	51,102
Annual cost*	US$/year	55,323	45,231	53,950	51,479
Annual savings	US$/year	-1,497	-325	1,769	-378
Electricity generated	US$/MWh	435	321	301	366
Payback Period	years	32	15	8	15

* With peak shaving; ** Without peak shaving

Table 12. Economical analysis of the peak shaving case

Figure 19 shows payback period in relation to microturbine cost. There is a strong fall on the payback period of the states of SP and RJ, due to a decrease in the microturbine cost.

A few manufactures intend to decrease microturbine costs to about 400 US$/kW until 2005 (Dunn & Flavin, 2000). If the microturbine cost is 400 US$/kW, the payback period will be between 2.5 and 5 years, as shown in Figure 19.

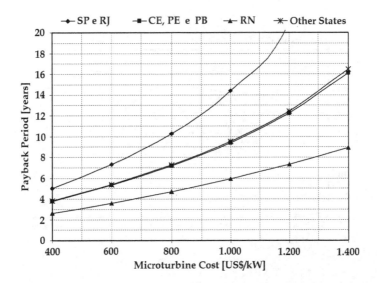

Figure 19. The influence of the microturbine cost on the return on investments.

8.2. Base load case

In this case, a microturbine produces electricity to a gas station according to the base load demand, as shows Figure 20. The conditions of this case are in table 13, whereas Table 14 shows the Brazilian regions analyzed in the base load case.

Model of microturbine	Capstone 330	
Number of microturbines	1	
Life time of microturbines	10	years
Net power	27,5	kW
Microturbine installed cost	1,538	US$/kW
Natural gas consumption (HHV)	6,918	m³/month
Average price of natural gas (taxes included)	0.24 - 1.02	R$/m³

Table 13. Conditions of the base load case.

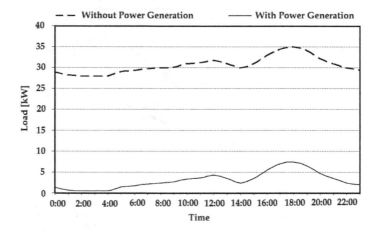

Figure 20. Electricity demand supplied by utility companies.

	Brazilian States
1st Region	São Paulo (SP) and Rio de Janeiro (RJ)
2nd Region	Rio Grande do Sul (RS) and Paraná (PR)
3rd Region	Rio Grande do Norte (RN)
4th Region	Others

Table 14. Brazilian regions analyzed in the base load case.

Table 15 displays the economical analysis of the base load case for gas stations. Up to the present moment this kind of business is not feasible, except in the state of Rio Grande do Norte (RN) where payback period can be 3.1 years, once local gas distribution companies have encouraged thermoelectric small scale power generation, according to natural gas price lower than others kind of fuels.

		SP e RJ	RS e PR	RN	Other States
Total Investment	US$	46827	46827	46827	46827
Annual cost**	US$/year	28748	28117	27956	24389
Annual cost*	US$/year	45699	33898	20707	26002
Annual savings	US$/year	- 16951	- 5780	7249	- 1614
Electricity generated	US$/MWh	181	131	75	99
Payback Period	years	Not Feasible	Not Feasible	3,1	8,2

* With power generation; ** Without power generation

Table 15. Economical analysis of the base load case.

Figure 21 shows the behavior of the cost of the electricity generated for different micro-turbine costs and natural gas average price. Some natural gas distribution companies in Brazil have encouraged the creation of small thermal power generation units, as the cost of natural gas coming from these companies would be about 0.24 R$/m^3. Based on this fact and on the perspective of microturbine manufactures, Figure 21 shows the cost of the electricity generated could be 58 US$/MWh. For each 1 US$/kW decreased from the microturbine cost, the cost of the electricity generated decreases about 0,021 US$/MWh, for every natural gas average price range, and for each 1 R$/m^3 decreased from the natural gas average price, the cost of the electricity generated decreases about 135 US$/MWh, for every microturbine cost range.

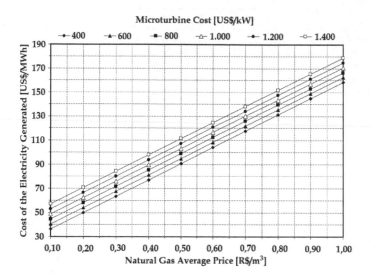

Figure 21. Cost of the electricity generated for different microturbine costs and natural gas average prices.

In the base load case, the natural gas average price is the most influential component in the return on investments. Figure 22 shows this conclusion for the microturbine cost at this moment, since natural gas average price of 0.24 R$/m³ can result in a payback period between 3 and 4 years.

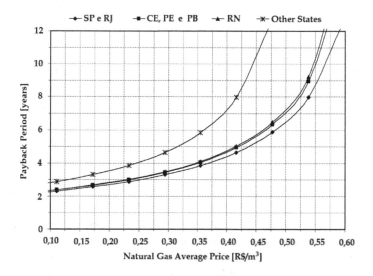

Figure 22. The natural gas average price influence on the payback period.

8.3. Cogeneration case

In this case, two microturbines and a heat recovery system produced electricity and hot water to buildings in a residential segment, according to the base load demand, as can be seen in Figure 23.

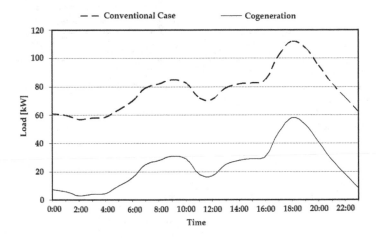

Figure 23. Electric demand supplied by utility companies to consumers with and without cogeneration.

A cogeneration plant can result in substantial savings of energy. However, these systems usually result in greater capital expenditures than non-cogeneration plants. This incremental capital investment for cogeneration must be justified by reduced annual energy costs and reduced payback periods.

A course of action involving minimum capital expenditures can be determined as the conventional case. In this study a low pressure boiler supplying process heat and the purchase of all electric power from utility system is the conventional case. Although the conventional case has the lowest investment cost, it usually has annual operating costs significantly higher than those available with cogeneration alternatives. Table 16 shows the conditions of this case, while Table 17 shows the Brazilian regions analyzed in the base load case.

System cogeneration model	MG2-C1	
Number of Capstone microturbines	2	
Number of heat recovery systems	1	
Life time of microturbines	10	years
Power output	54	kW
Heat recovery systems (hot water generation)		
Water pressure	10	bar
Water flow	2.22	t/h
Inlet water temperature	25	°C
Outlet water temperature	67	°C
Outlet exit gas temperature	93	°C
Net power	53	kW
System cogeneration installed cost	1,872	US$/kW
Natural gas consumption (HHV)	13,653	m³/day
Average price of natural gas (taxes included)	0.24 - 0.90	R$/m³

Table 16. Conditions of the cogeneration case.

	Brazilian States
1st Region	Rio de Janeiro (RJ)
2nd Region	Paraná (PR)
3rd Region	Rio Grande do Norte (RN)
4th Region	Others

Table 17. Brazilian regions analyzed in the cogeneration case.

Table 18 displays the economical analysis of the cogeneration case. Investments costs are lower in the conventional case than in the cogeneration system, and, although the annual cost is higher, savings can be up to US$ 24,907 per year. The payback period is between 2.8 and 3.8 years and the minimal cost of the electricity generated is 84 US$/MWh.

		RJ	PR	RN	Other States
Total Investment*	US$	23077	23077	23077	23077
Total Investment**	US$	136797	136797	136797	136797
Annual cost*	US$/year	128566	110022	90323	98328
Annual cost**	US$/year	110337	96655	65416	77534
Annual savings	US$/year	18228	13367	24907	20795
Electricity generated	US$/MWh	174	146	84	112
Payback Period	years	3,3	3,8	2,8	3,1

* Conventional; ** Cogeneration

Table 18. Economical analysis of the cogeneration case.

In the cogeneration case, the fuel cost is the most influential component on the return on investment, similar to the base load case. Figure 24 shows fuel costs can represent up to 71% of the cost of the electricity generated.

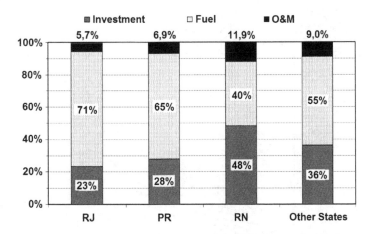

Figure 24. Components of the cost of the electricity generated.

Figure 25, Figure 26 and Figure 27 show the combined influence of microturbine cost and the average price of natural gas on the return on investment in the states of Rio de Janeiro and Paraná (Figure 25), Rio Grande do Norte (Figure 26) and the other states (Figure 27). Based on the perspective of microturbine manufactures and with natural gas average price of 0.25 R\$/m³, the payback period can be between 1.5 and 3 years.

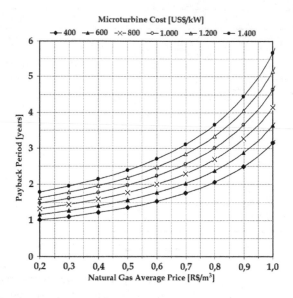

Figure 25. Combined influence of microturbine cost and average price of natural gas on the payback period in the states of Rio de Janeiro and Paraná.

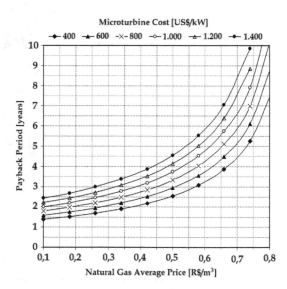

Figure 26. Combined influence of microturbine cost and average price of natural gas on the payback period in the state of Rio Grande do Norte.

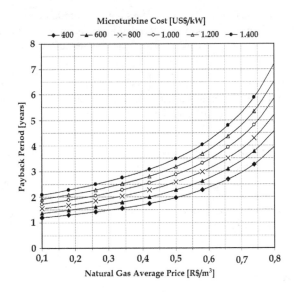

Figure 27. Combined influence of microturbine cost and average price of natural gas on the payback period in the other states.

9. Conclusions

The variable speed operation and the electric power conditioner increase part-load efficiency of microturbines as they allow for the improvement of part-load fuel savings, especially increased recuperator effectiveness at lower part-load airflows. The variable speed control improves part-load performance but requires a system able to sense load and optimize speed. According to the results shown in this study, the microturbines efficiency is at its highest when Capstone microturbines are operating over an output range between 12 kW and full load.

Capstone microturbines use clean combustion technology to achieve low emissions. Nitrogen oxides (NO_x) and carbon monoxide (CO) emission levels of these machines are lower than 7 ppmv@15%O_2 at full load when these microturbines are fueled with natural gas.

Microturbines exhibit low emissions of all classes of pollutants and have environmental benefits as they release fewer emissions compared to other distributed generation technologies, like internal combustion engines. Besides, these units are clean enough to be placed in a community with residential and commercial buildings.

Microturbine generators have shown good perspectives for electricity distributed generation in small scales, once they have high reliability and simple design (high potential for large scale cheap manufacturing).

Although results show microturbines are not feasible to provide energy at peak demand, in this case the microturbines can supply peak demand and improve the level of reliability of the electricity supplying, because they can provide stand-by capabilities should the electric grid fail.

In the base load case this sort of business is feasible just in states of Brazil where natural gas distributing companies have encouraged small thermal power generation by natural gas with lower prices, since the price is the most influential cost component of the electricity generated.

The most feasible investment in microturbines is in the cogeneration case. In this case, economical feasibility is certain in all states of Brazil as cogeneration systems can provide considerable annual savings. Besides, under the perspective of manufacturers, and with the incentive of natural gas distribution companies together with the rise in electricity prices of Brazilian utility companies, investments in microturbines for the next years will be higher than currently.

Acknowledgements

The authors would like to thank CAPES, FAPEMIG, FAPEPE and CNPq, for their financial support.

Author details

Marco Antônio Rosa do Nascimento, Lucilene de Oliveira Rodrigues,
Eraldo Cruz dos Santos, Eli Eber Batista Gomes, Fagner Luis Goulart Dias,
Elkin Iván Gutiérrez Velásques and Rubén Alexis Miranda Carrillo

Federal University of Itajubá – UNIFEI, Brazil

References

[1] Asme performance test code PTC-22-1997, Gas turbine power plants, 1997.

[2] Barker, T. Micros, Catalysts and Electronics, Power-Gen International 96, Turbomachinery, v. 38, n. 1, p. 19-21, 1997.

[3] Biasi, V. de Low cost and high efficiency make 30 to 80 kW microturbines attractive, Gas Turbine World, Jan.-Feb., Southport, 1998.

[4] Bolszo, C. D., and Mcdonell, V. G., Emissions Optmization of a Biodiesel Fired Gas Turbine, Proceedings of the Combustion Institute, 32, LSEVIER, pp. 2949-2956, 2009.

[5] Capstone Turbine Corporation, Capstone Low Emissions Microturbine Tecnology, White Paper, USA, 2000.

[6] Capstone Turbine Corporation, Capstone Microturbine Model 330 System Operation Manual, USA, 2001.

[7] Capstone Turbine Corporation, Capstone Microturbine Product Catalog, USA, 2012: http://www.capstoneturbine.com/prodsol/products/, accessed at: 20/06/2012.

[8] Cohen, H., Rogers, G. F. C., and Saravanamuttoo, H. I. H., Gas Turbine Theory, Fourth edition, 1996.

[9] Dunn, S. & Flavin, C., Dimensionando a Microenergia. In: Estado do Mundo 2000. Brazil, UMA Ed., 2000.

[10] Gomes, E. E. B. Análise Técnico-econômica e Experimental de Microturbinas a Gás Operando com Gás Natural e Óleo Diesel, Master Degree Thesis, Supervised by Nascimento, M. A. R. and Lora, E. E. S. Federal University of Itajubá, 2002.

[11] Hamilton, S. L., Microturbines, Distributed Generation: a nontechnical guide, edited by Ann Chambers, cap. 3, pp. 33 – 72, PennWell Corporation, USA, 2001

[12] GRI - Gas Research Institute, The role of Distributed Generation in competitive energy markets, Distributed Generation Forum, Gas Research Institute (GRI), 1999.

[13] Liss, W.E., Natural Gas Power Systems for the Distributed Generation Market. Power-Gen International '99 Conference. CD-Rom. New Orleans, Louisiana, USA, 1999.

[14] Nascimento, M. A. R.; Santos, E. C., Biofuel and Gas Turbine Engines, Advances in Gas Turbine Technology, InTech, ISBN 978-953-307-611-9, chaper 6, 2011.

[15] Nwafor, O., Emission characteristics of Diesel engine operating on rapeseed methyl ester. Renewable Energy, 29, pp. 119-29, 2004.

[16] Pierce, J. L. Microturbine Distributed Generation Using Conventional and Waste Fuel, Cogeneration and On-Site Power Production, James & James Science Publishers, p. 45, v. 3, Issue 1, Jan-Feb, 2002.

[17] Rodgers, C.; Watts, J.; Thoren, D.; Nichols, K. & Brent, R. Microturbines, Distributed Generation – The Power Paradigm for the New Millennium, edited by Anne-Marie Borbely & Jan F. Kreider, cap. 5, pp. 120 – 148, CRC Press LLC. USA, 2001a.

[18] Rodgers, G., and Saravanamutto, H., Gas Turbine Theory, Prentice Hall, 2001b.

[19] Scott, W. G. Micro Gas Turbine Cogeneration Applications, International Power and Light Co., USA, 2000.

[20] Sierra, R. G. A., Teste Experimental e Análise Técnico-Econômica do Uso de Biocombustíveis em uma Microturbina a Gás de Tipo Regenerativo; Dissertação de Mestrado, UNIFEI, 2008.

[21] Watts, J. H, Microturbines: a new class of gas turbine engine, Global Gas turbine News, ASME-IGTI, v. 39, n. 1, p. 4-8, USA, 1999.

[22] Weston, F., Seidman, N., L., James, C. Model Regulations for the Output of Specified Air Emissions from Smaller-Scale Electric Generation Resources, The Regulatory Assistance Project, 2001.

Chapter 3

On Intercooled Turbofan Engines

Konstantinos G. Kyprianidis, Andrew M. Rolt and
Vishal Sethi

Additional information is available at the end of the chapter

1. Introduction

Public awareness and political concern over the environmental impact of the growth in civil aviation over the past 30 years have intensified industry efforts to address CO_2 emissions [5]. CO_2 emissions are directly proportional to aircraft fuel burn and one way to minimise the latter is by having engines with reduced Specific Fuel Consumption (SFC) and installations that minimise nacelle drag and weight. Significant factors affecting SFC are propulsive efficiency and thermal efficiency. Propulsive efficiency has been improved by designing turbofan engines with bigger fans to give lower specific thrust (net thrust divided by fan inlet mass flow) until increased engine weight and nacelle drag have started to outweigh the benefits. Thermal efficiency has been improved mainly by increasing the Overall Pressure Ratio (OPR) and Turbine Entry Temperature (TET) to the extent possible with new materials and design technologies.

Mission fuel burn benefits from reducing specific thrust are illustrated in Fig. 1 (for a year 2020 entry into service, but otherwise conventional, direct drive fan engine for long range applications). The engine Take-Off (TO) thrust at Sea Level Static International Standard Atmosphere (SLS ISA) conditions is 293.6kN (66000lbf) and all Fan Pressure Ratio (FPR) and ByPass Ratio (BPR) values quoted are at mid-cruise conditions. The figure shows that only a modest reduction in block fuel is obtained by increasing the already large fan diameter. Reduced powerplant weight and/or nacelle drag would be needed before lower specific thrust would be justified, and one way of doing this would be to discard the nacelle and fit an open rotor in place of the fan.

An alternative design approach to improving SFC is to consider an increased OPR intercooled core performance cycle. The thermal efficiency benefits from intercooling have been well documented in the literature - see for example [2, 3, 7, 9, 11–13, 15]. Very little information is available however, with respect to design space exploration and optimisation for minimum block fuel at aircraft system level.

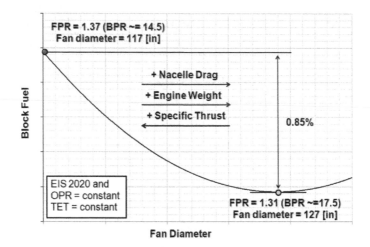

Figure 1. Block fuel benefits from reducing specific thrust for a year 2020 entry into service conventional turbofan engine for long range applications.

Previously, a comparative study was presented focusing on a conventional core and an intercooled core turbofan engine for long range applications [5, 7]. Both configurations had the same fan diameter and were designed to meet the same thrust requirements. They were Ultra-High Bypass Ratio (UHBR) designs based on a three-shaft layout with a direct drive front fan. The intercooled core configuration (illustrated in Fig. 2) featured an intercooler mounted inboard of the bypass duct. The installation standard included a flow splitter and an auxiliary variable geometry nozzle. The two concepts were evaluated based on their potential to reduce CO_2 emissions (and hence block fuel) through both thermal and propulsive efficiency improvements, for engine designs to enter service between 2020 and 2025. Although fuel optimal designs were proposed, only limited attention was given to the effect of design constraints, material technology and customer requirements on optimal concept selection.

A study is presented here that focuses on the re-optimization of those same powerplants by allowing the specific thrust (and hence the propulsive efficiency) to vary. Rather than setting fixed thrust requirements, a rubberised-wing aircraft model was fully utilised instead. The engine/aircraft combination was optimized to meet a particular set of customer requirements i.e. payload-range, take-off distance, time to height and time between overhaul. It was envisaged that different conclusions would be drawn when comparing the two powerplants at their optimal specific thrust and absolute thrust levels. It is shown through this study that performing a comparison at each concept's optimal specific thrust level gives a different picture on intercooling. Differences in the optimal specific thrust levels between the two configurations are discussed. The design space around the proposed fuel-optimal designs was explored in detail and significant conclusions are drawn.

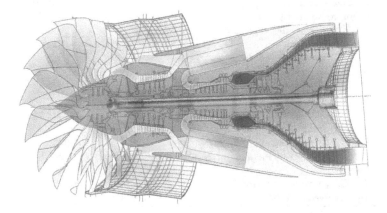

Figure 2. Artistic impression of the intercooled core turbofan engine [10].

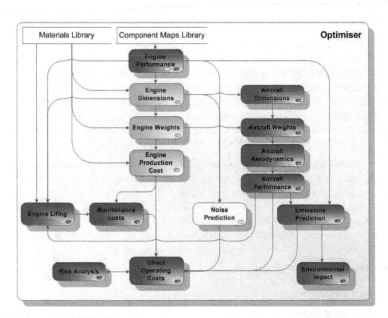

Figure 3. Conceptual design tool algorithm [4].

	Lower bound	Upper bound
FAR take-off distance	-	2.5 [km]
Climb to 35000 [ft]	-	22.5 [min]
IPC design pressure ratio (intercooled core)	2.7	-
HPC design pressure ratio (intercooled core)	-	25.0
HPC design pressure ratio (conventional core)	-	5.5
HPC delivery temperature	-	970 [K]
HPC last stage blade height	10 [mm]	-
Combustor outlet temperature	-	2050 [K]
Turbine blade mean metal temperature (external surface)	-	1350 [K]
Auxiliary nozzle area variation	Ref.	+50%
Time between overhaul	23000 [hr]	-

Table 1. Design space constraints.

2. Methodology

To effectively explore the design space a tool is required that can consider the main disciplines typically encountered in conceptual design. The prediction of engine performance, aircraft design and performance, direct operating costs and emissions for the concepts analysed in this study was made using the code described in [6]. Another code described in [7], was also used for carrying out the mechanical and aerodynamic design in order to derive engine component weights and dimensions. The two tools have been integrated together within an optimizer environment as illustrated in Fig. 3 with a large amount of information being made available to the user during the design iteration. The integration allows for multi-objective optimization, design studies, parametric studies, and sensitivity analysis. In order to speed up the execution of individual engine designs, the conceptual design tool minimizes internal iterations in the calculation sequence through the use of an explicit algorithm, as described in detail by Kyprianidis [4].

For every engine design there are numerous practical limitations that need to be considered. A comprehensive discussion on design constraints for low specific thrust turbofans featuring conventional and heat exchanged cores can be found in [5]. The design space constraints set for this study are given in Table 1 and are considered applicable to a year 2020 entry into service turbofan engine. The effect on optimal concept selection of design constraints, material technology and customer requirements is discussed in the following sections.

3. Optimising a turbofan engine

3.1. Fuel-optimal designs

Optimizing a turbofan engine design for minimum block fuel essentially has to consider the trade-off between better thermal and propulsive efficiency and reduced engine weight and nacelle drag. The cycle optimization results for the two powerplants are given in Table 2.

	Conventional core EIS 2020	Intercooled core EIS 2020
Fan diameter [in]	127	121
ISA SLS take-off thrust [lbf]	66000	64500
Overall pressure ratio	62.3	80.2
IPC pressure ratio	8.0	3.8
HPC pressure ratio	5.5	15.5
Fan mass flow [kg/s]	588	525
Core mass flow [kg/s]	36.3	34.6
Mid-cruise fan tip pressure ratio	1.30	1.39
Mid-cruise bypass ratio	17.7	17.3
Mid-cruise SFC	Ref.	-1.5%
Mid-cruise thermal efficiency (core + transmission efficiency)	Ref.	+0.019
Mid-cruise propulsive efficiency	Ref.	-0.021
Engine installed weight	Ref.	-11.0%
Fan weight	Ref.	-21.3%
LPT weight	Ref.	-25.6%
Core weight	Ref.	-20.9%
Added components weight (as % of engine dry weight)	-	10.5%
Nacelle weight	Ref.	-14.7%
MTOW [1000 kg]	208.5	203.4
OEW [1000 kg]	116.2	113.1
Block fuel weight	Ref.	-3.0%

*Performance parameters at top of climb conditions unless stated otherwise

Table 2. Comparison of the fuel optimal intercooled and conventional core turbofan engine designs.

Significant block fuel benefits are projected for the intercooled core engine, but they are smaller than those predicted in previous efforts [7]. This is mainly attributed to a minimum blade height requirement setting a practical lower limit on the intercooled core size for a given OPR. Increasing the fan diameter at a fixed tip speed inevitably reduces rotational speed, increases torque and hence increases the Low Pressure (LP) shaft diameter; this further aggravates the problem since the High Pressure Compressor (HPC) hub to tip ratio needs to increase. As a result, the optimal specific thrust for the intercooled core is higher compared to the conventional core turbofan engine. Although the high OPR intercooled core benefits from a higher core and transmission efficiency, and hence a better thermal efficiency, the conventional core benefits from a higher propulsive efficiency. The design space around the proposed fuel optimal designs was explored and in the next sections important observations are presented.

3.2. Approximating the design space

In order to graphically illustrate the design space, a large number of simulations had to be carried out; these simulations were focused around the fuel-optimal designs presented in Section 3.1. Polynomial response surface models were derived that interpolate between

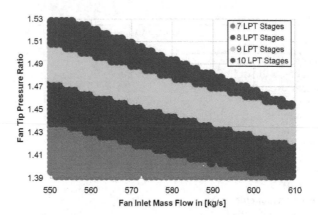

Figure 4. Variation of low pressure turbine stage count with fan inlet mass flow and fan tip pressure ratio for a fixed size conventional core.

a given number of known designs. Typical design space discontinuities encountered as a result of turbomachinery stage count changes are inevitably distorted in polynomial approximations. For this reason, an error analysis was carried out to determine the discrepancy levels between the surrogate models and the actual design spaces; the approximation errors for engine weight and aircraft block fuel were found to be less than 1% and 0.2%, respectively.

3.3. Fan and core sizing

Propulsive efficiency benefits from reducing specific thrust by increasing fan diameter can very well be negated by the resulting combination of: i) increased engine and nacelle weight, ii) increased nacelle (and interference) drag, and iii) reduced transmission efficiency. This section discusses various aspects of fan and core sizing for the conventional core and intercooled core configurations.

When sizing the engine fan, assuming a fixed size core (i.e., fixed core inlet mass flow), large design space discontinuities are encountered due to Low Pressure Turbine (LPT) stage count changes, as illustrated in Fig.4.

As discussed earlier, the use of smooth surrogate models for approximating discontinuous spaces inevitably results in approximation errors, and it is worth noting that the addition of an extra LPT stage results in approximately 150kg of additional weight. Nevertheless, with the fan and nacelle weight (including the thrust reverser) each being roughly double the LPT weight and directly proportional to the fan diameter, the weight trends illustrated in Fig. 5 can be considered reasonable.

The improvement in mid-cruise uninstalled SFC from reducing specific thrust is illustrated in Fig. 6. If installation effects are ignored, then selecting a higher fan diameter (and hence a higher bypass ratio at a fixed size core) will result in better SFC. Nevertheless, the increased nacelle drag and engine weight move the optimal level of specific thrust for minimum block fuel to smaller fan diameters, as illustrated in Fig. 7.

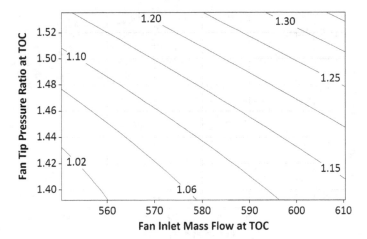

Figure 5. Variation of engine weight with fan inlet mass flow and fan tip pressure ratio for a fixed size conventional core.

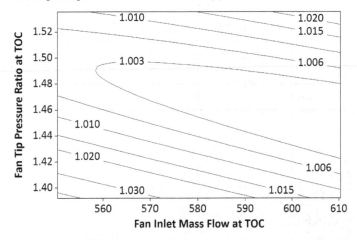

Figure 6. Variation of engine specific fuel consumption with fan inlet mass flow and fan tip pressure ratio for a fixed size conventional core.

Looking at the trends illustrated in Fig. 7 in isolation, and then comparing with the optimal design proposed in Section 3.1, one would be inclined to draw the conclusion that the fuel-optimal fan diameter should be even smaller. However, as one moves towards the upper left corner of Fig. 7 the engine take-off and Top Of Climb (TOC) thrusts reduce (because the core size is fixed and the fan is getting smaller. In order to satisfy - at constant specific thrust - the time to height and FAR (Federal Aviation Regulations) take-off distance constraints set in this study it is necessary to scale-up the engine i.e., increase fan and core size simultaneously which leads to: i) higher engine and nacelle weight, ii) higher nacelle drag, and iii) non-optimum engine/aircraft matching i.e. mid-cruise conditions are away from the bottom of the SFC loop (particularly if other cycle parameters are not re-optimized).

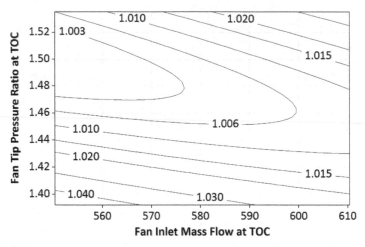

Figure 7. Variation of aircraft block fuel with fan inlet mass flow and fan tip pressure ratio for a fixed size conventional core.

Most of the conclusions drawn in this section are applicable to both the conventional core and the intercooled core configurations. Nevertheless, the intercooled core is constrained by a practical minimum blade height requirement for the last HPC stage (assuming an all-axial bladed HPC). At a fixed core OPR and intercooler effectiveness, this constraint sets a minimum limit for the core mass flow and as a consequence a minimum limit is also set on specific thrust at a fixed engine thrust. This makes the intercooled core more favourable for very high thrust engines, as they will not be subject to this constraint.

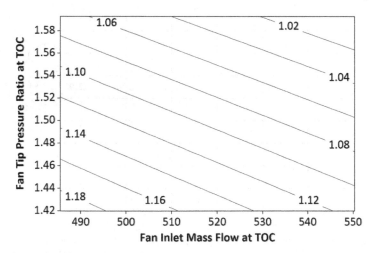

Figure 8. Variation of HPC last stage blade height with fan inlet mass flow and fan tip pressure ratio for a fixed size intercooled core.

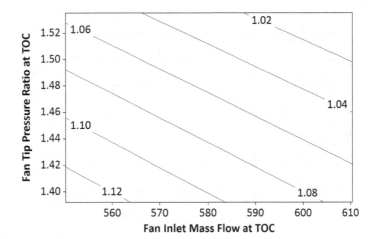

Figure 9. Variation of HPC last stage blade height with fan inlet mass flow and fan tip pressure ratio for a fixed size conventional core.

Bigger direct drive fans rotating at low speeds result in high torque requirements which increase the LP shaft outer diameter. The HPC inner diameter has to be pushed out and therefore slowed down, so for a given flow area and blade speed, the resulting blade height tends to reduce, as illustrated in Fig. 8 - the problem is less marked for a conventional core as illustrated in Fig. 9. For a given blade height requirement the core mass flow needs to be increased and it can therefore be concluded that an intercooled core would favour a geared fan arrangement, over a direct drive one, since it could alleviate some of the restrictions set on the cycle. An aft fan arrangement as the one presented in [1] could further relieve this issue by not passing the LP shaft through the core, though aft fan arrangements set other design challenges.

3.4. IPC/HPC work split

Increasing engine OPR improves thermal efficiency and hence SFC, as illustrated in Fig. 10. The optimal OPR level for the conventional core is constrained by the maximum allowable HPC delivery temperature set, as illustrated in Fig. 11. For the intercooled cycle, this limitation is alleviated but only to be replaced by a practical minimum blade height requirement which consequently sets a minimum allowable core size constraint. The optimal OPR level for the intercooled core at a fixed specific thrust is therefore a trade-off between a better core efficiency and a smaller core size.

If one assumed constant component polytropic efficiencies then SFC benefits would arise for the conventional core from shifting pressure ratio to the more efficient High Pressure (HP) spool.

However, as the HPC pressure ratio rises beyond an upper limit set, the core configuration would inevitably need to be changed to a two-stage High Pressure Turbine (HPT). This would introduce higher cooling flow requirements (and hence losses) and could also make the core

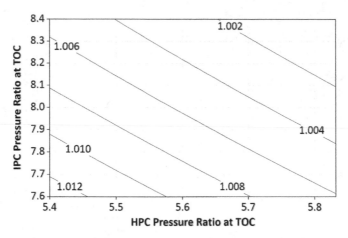

Figure 10. Variation of mid-cruise specific fuel consumption with IPC and HPC pressure ratio for a fixed size conventional core.

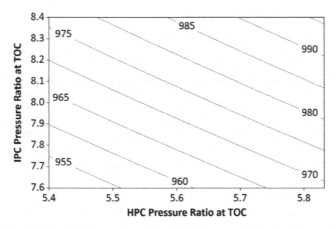

Figure 11. Variation of take-off HPC exit temperature with IPC and HPC pressure ratio for a fixed size conventional core.

heavier and longer, negating the originally projected benefits. Efficient intercooling requires that the Intermediate Pressure Compressor (IPC) has significantly less pressure ratio than the HPC [14]. For that reason, a two-stage HPT has been assumed for the intercooled core while a minimum IPC design pressure ratio was set to avoid potential icing problems during decent.

3.5. Engine ratings

Sizing and rating an engine is a highly complex process that has to consider aircraft performance requirements, fuel consumption, and engine lifing. Turbine blade lifing requirements and cooling technology set a maximum allowable blade metal temperature

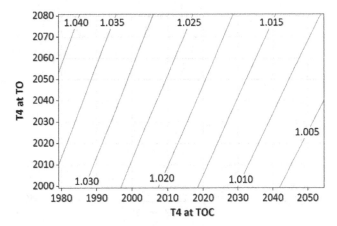

Figure 12. Variation of engine weight with combustor outlet temperature at take-off and top of climb conditions for a fixed size conventional core.

constraint; cooling flows therefore need to increase with increasing combustor outlet temperature (T_4) levels. The maximum T_4 level may also be constrained by combustor design considerations. For example increasing combustor liner cooling requirements essentially reduces the amount of air available for mixing in the combustion zone and hence the flame temperatures and NO_x emissions tend to increase. Detail design studies are required for establishing the optimal trade-off between cycle efficiency and acceptable NO_x levels. For these reasons an upper limit was set for T_4 that was considered to be a reasonable trade-off for year 2020 entry into service turbofan engines. The same limit was used for both the conventional core and the intercooled core.

Although the intercooled core benefits from lower combustor inlet temperatures, the air to fuel ratio is lower for a given T_4. Furthermore, high pressure levels in the intercooled cycle will affect the influence of luminosity on gas emissivity, and hence the temperature difference across the liner [8].

For a given OPR there is an optimal mid-cruise T_4 for best SFC. Nevertheless, running the cycle hotter at top of climb (than the optimal for mid-cruise SFC) tends to reduce engine weight, as illustrated in Fig. 12. These benefits come mainly from the reduction in LPT weight since a higher T_4 results in a more efficient core expansion and hence a higher pressure and lower corrected mass flow at the LPT inlet. A further reduction in weight is possible through the reducing core size (mainly in the case of the conventional core) since core output is increasing with T_4. On the other hand, running the cycle hotter at hot day take-off can lead to an increase in engine weight at a fixed core size. An increase in T_4 at top of climb generally requires an increase in T_4 at take-off in order to maintain a constant FAR take-off field length. T_4 at top of climb is therefore constrained by a hot-day take-off T_4 limitation. Furthermore, with modern large engines on long range aircraft typically being heavily derated at take-off conditions milder than hot-day and/or with less than a full fuel load, top of climb T_4 will want to be lower than hot-day take-off T_4 so as not to compromise engine life [12]. An optimal block fuel trade-off therefore arises as illustrated in Fig. 13.

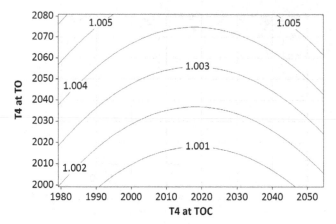

Figure 13. Variation of aircraft block fuel with combustor outlet temperature at take-off and top of climb conditions for a fixed size conventional core.

3.6. Intercooler effectiveness

In this study the aerodynamic design for most engine components has been carried out at top of climb conditions. However, the intercooler component has been sized at end of runway hot day take-off conditions (kink point) were the highest heat transfer levelswhere the highest heat transfer levels are encountered. At cruise conditions the variable geometry dual-nozzle system is utilised to reduce the intercooler mass flow ratio (intercooler cooling mass flow divided by core mass flow) and hence reduce intercooler cold side pressure losses. This practice results in better SFC and hence lower block fuel.

Engine design variations focused around the fuel optimal design are presented in Fig. 14 in a similar manner to figures presented in earlier sections. The figure illustrates the effect of intercooler effectiveness on weight. As can be observed, intercooler effectiveness at top of climb conditions has only a second order effect on intercooler weight while intercooler effectiveness at take-off conditions has a more significant effect. As intercooler weight increases, so does block fuel. Further to increasing intercooler weight, increasing intercooler cooling air flow and effectiveness at take-off conditions increases thrust at a given combustor outlet temperature. This thrust improvement however is soon negated by increasing intercooler cold side pressure losses, as discussed in detail in [7].

It can be observed in Fig. 15 that there is a limit to the block fuel benefit that may be achieved by optimising the intercooler effectiveness level at different flight conditions. This limitation is set by: i) the maximum allowable nozzle area variation (dot-dashed white iso-lines), and from ii) the reducing overall pressure ratio level during cruise conditions (white continuous iso-lines). Although at first glance it seems to be implied through this figure that a low intercooler effectiveness is beneficial for block fuel, it should be noted that a minimum level of intercooler effectiveness has to be maintained at take-off (and hence at cruise due to the aforementioned nozzle area variation limitation). This is due to the need for satisfying a maximum FAR take-off field length requirement at a given maximum combustor outlet temperature. An optimal trade-off therefore exists between intercooler effectiveness,

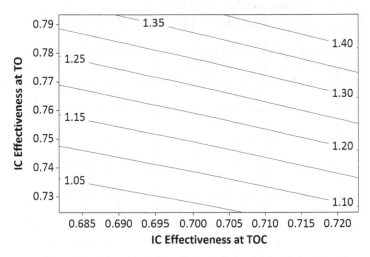

Figure 14. Variation of intercooler weight with intercooler effectiveness at top of climb and take-off conditions.

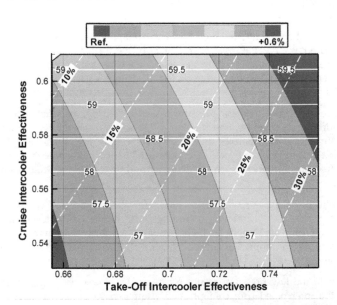

Figure 15. Variation of block fuel with intercooler effectiveness at take-off and cruise conditions.

specific thrust, core size, and overall pressure ratio. It should be stressed that the optimum design intercooler effectiveness level also clearly depends on the heat exchanger technology available.

These significant performance benefits - being the result of controlling the amount of cooling flow going through the intercooler, and hence the effectiveness and pressure loss levels at different operating points - may be achieved not only by utilising a variable geometry dual-nozzle system but alternatively through a variable area mixer, which returns spent intercooler air to the bypass duct. Optimal variable geometry settings can be identified for different operating points and the projected benefits are up to 2% increase in net thrust (F_N) at take-off and 2% reduction in SFC at cruise.

4. Sensitivity analysis of optimal designs

The work presented in this section aims to deliver averaged exchange rates which can be used to investigate the effect of technology parameter deviations on block fuel. Information on how these perturbations were introduced in the design algorithm is given in Appendix A.

The sensitivity parameters compiled allow for system level quantification of the importance of research on specific component technologies i.e. they can be used to assess the significance of progress in specific component technologies for each engine configuration. Inversely, these exchange factors also help quantify the impact of technology shortfalls. The exchange rates presented in Fig. 16 and Fig. 17 should be perceived as fractional percentage variations from the technology target values that were assumed when deriving the fuel optimal designs presented in Section 3.1.

For the conventional core configuration for long range applications the low pressure system component technology has the greatest influence on performance, as expected for a low specific thrust engine. Significant fuel burn benefits are expected by improving fan and LPT efficiency. Inversely, shortfalls in meeting projected technology targets for the low pressure system will have a major impact on overall engine/aircraft performance.

As fan tip pressure ratio reduces, pressure losses in the bypass duct tend to have an increasingly dominant effect on transmission efficiency and, therefore, on the impact of propulsive efficiency improvements on SFC. By combining Fig. 1 and Fig. 16 it can be observed that a 10% increase in bypass duct pressure losses would halve the projected block fuel benefits from a 10 [in] increase in fan diameter and the consequent reduction in specific thrust.

Failure to deliver the expected efficiency levels for the compressor components would increase combustor inlet temperatures resulting in higher NO_x levels and reduced component life. Combustor designs are highly sensitive to inlet conditions and it is likely that a significant shortfall in compressor efficiency would require a re-design of the combustor as well as the compressors.

The influence of the low pressure system component technology on performance is less marked for the intercooled core configuration compared to the conventional core. The difference in the exchange rates is directly proportional to the difference in specific thrust between the two optimal designs.

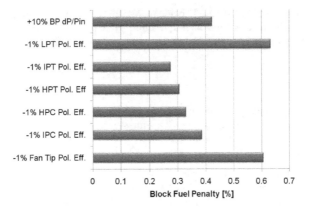

Figure 16. Sensitivity analysis around the fuel optimal design for the conventional core configuration.

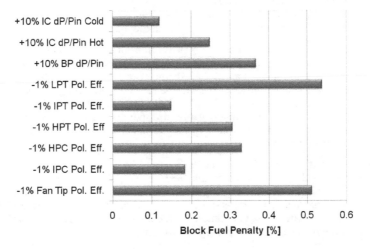

Figure 17. Sensitivity analysis around the fuel optimal design for the intercooled core configuration.

The efficiencies of the IPC and Intermediate Pressure Turbine (IPT) in the intercooled core for long range applications have a significantly smaller influence on block fuel, compared to the conventional core configuration, which reflects the significantly lower pressure ratio on the Intermediate Pressure (IP) spool. On the other hand, the efficiencies of the HPC and HPT have a similar influence on block fuel, compared to the conventional core configuration, despite the significantly higher pressure ratio placed on the HP spool. This can be explained by the fact that by reducing the HPC inlet temperature, intercooling significantly reduces HP compression work at a pressure ratio, and also increases the specific power of the core.

As can be observed, intercooler pressure losses have a significant effect on block fuel. Losses in the intercooler hot stream are more significant than losses in the cold stream at cruise and climb, while losses in the cold stream become increasingly important as the intercooler mass flow ratio (W132Q25) increases at take-off. Failure to achieve the intercooler pressure loss targets set could significantly reduce the projected block fuel benefits for the intercooled core configuration.

5. Conclusions

In this study, the combined potential of novel low pressure spool and core technologies was assessed with respect to reducing engine CO_2 emissions. A back-to-back comparison of an intercooled core engine with a conventional core engine was performed and fuel optimal designs for year 2020 entry into service were proposed.

The results from the optimization process show that the optimal specific thrust for the intercooled core is somewhat higher compared to the conventional core turbofan engine. This is mainly attributed to the HPC last stage blade height requirement limiting minimum core size in the intercooled engine and negating one of the benefits of increasing fan diameter. This conclusion may appear specific to the thrust scale of the study engine and it might not apply to more powerful engines, but it is considered likely to be generally applicable because all intercooled engines have relatively small core size and so will be more susceptible to the loss of core component efficiency associated with making the core smaller still.

The optimized high OPR intercooled core benefits from higher thermal efficiency, but the optimized conventional core still benefits from higher propulsive efficiency. As a remedy to this, it is proposed to remove the LP shaft diameter constraint to enable the intercooled engine to have a faster more efficient lower hub to tip ratio core. This may be achieved by having a geared fan and a high-speed LP turbine with a smaller diameter shaft, or an aft fan arrangement (with a geared or counter-rotating turbine) or by having a reverse-flow core. Any of these arrangements might reduce the optimal specific thrust level significantly but would make 2020 a very ambitious target for entry into service.

It can be concluded that significant benefits in terms of block fuel are possible from an intercooled core, with year 2020 entry into service level of technology, compared to a conventional core turbofan engine for long range applications. *However, the benefits are highly dependent on achieving technology targets such as low weight and pressure losses for the intercooler.* The commercial competitiveness of an intercooled core turbofan design will largely depend on how the aviation market evolves in the years to come.

Acknowledgements

This study has been performed under the project NEWAC (European Commission Contract No. AIP5-CT-2006-030876). The authors gratefully acknowledge this funding as well as the project partners collaboration. In more detail, the work in this paper was performed under NEWAC WP1.3, "Techno-Economic and Environmental Risk Assessment" and Cranfield University, and Rolls-Royce plc specifically contributed to the work presented in the paper. The authors are grateful to J.A. Borradaile, S. Donnerhack (MTU Aero Engines), A. Lundbladh (GKN Aerospace), T. Grönstedt (Chalmers University), L. Xu (Siemens), B. Lehmayr (University of Stuttgart) and A. Alexiou (National Technical University of Athens) for the stimulating discussions on advanced concepts and aero engine design. Many thanks go to the reviewers of this work for their constructive suggestions to improve the overall quality and clarity of the article.

Nomenclature

BP	ByPass duct
BPR	ByPass Ratio
CO_2	Carbon dioxide
dP/P_{in}	Fractional pressure loss
EIS	Entry Into Service
F_N	Net thrust
FAR	Federal Aviation Regulations
FL	Flight Level
FPR	Fan Pressure Ratio
HP	High Pressure
HPC	High Pressure Compressor
HPT	High Pressure Turbine
IC	InterCooler
ICAO	International Civil Aviation Organisation
IP	Intermediate Pressure
IPC	Intermediate Pressure Compressor
IPT	Intermediate Pressure Turbine
ISA	International Standard Atmosphere
LP	Low Pressure
LPT	Low Pressure Turbine
Mid-Cr	Mid-Cruise
MTOW	Maximum Take-Off Weight
NO_x	Nitrogen oxides
OEW	Operating Empty Weight
OPR	Engine Overall Pressure Ratio
P_2	Fan inlet pressure
P_{23}	Intermediate pressure compressor inlet pressure
P_{25}	Intermediate pressure compressor outlet pressure
P_3	High pressure compressor outlet pressure
Pol. Eff.	Polytropic Efficiency
PR	Pressure Ratio
SFC	Engine Specific Fuel Consumption
SLS	Sea Level Static
T_4	Combustor outlet temperature
T_{41}	High pressure turbine rotor inlet temperature
TET	Turbine Entry Temperature
TO	Take-Off
TOC	Top Of Climb
UHBR	Ultra High Bypass Ratio

Appendix: Optimisation design variables

This appendix provides additional information on the choice of design variables for the optimisation process utilised in this article. Unless explicitly stated otherwise, design variables refer to top of climb engine operating conditions (ISA +10 [K], FL350, Mach=0.82) which is set as the reference (design) point for engine performance. The effect of introducing a single design variable perturbation on the values of other parameters at design point and off-design conditions is described by Fig. 18 and Fig. 19, respectively. Similarly, Fig. 20 describes the effect of such perturbations on the values of mechanical design parameters and objective functions.

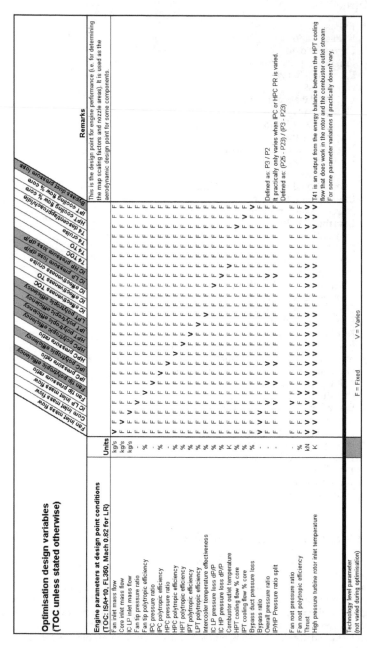

Figure 18. Main design variables for the direct drive fan conventional and intercooled core engine configuration and their effect on the values of other parameters at top of climb.

Optimisation design variables (TOC unless stated otherwise)	Units	Core inlet mass flow	IC LP inlet mass flow	Fan inlet mass flow	Fan tip pressure ratio	IPC pressure ratio	HPC polytropic efficiency	HPC pressure ratio	IPT polytropic efficiency	LPT polytropic efficiency	IC effectiveness TOC	IC effectiveness cruise	IC HP pressure loss dP/P	IC LP pressure loss dP/P	T4 TOC	T4 TO	T4 cruise	HPT des cent/approach/idle	IPT cooling flow % core	Bypass duct pressure loss	Remarks
Engine parameters at different off-design conditions (T/O: ISA+15, FL0, Mach 0.25; ICAO T/O: ISA SLS; Mid-Cr: ISA, FL350, Mach 0.82)																					
Take-off thrust	kN	V	V	V	V	V	V	V	V	V	V	V	F	F	V	F	F	V	V	V	This is the end of runway hot day TO (kink) point. It is used as the mechanical design point (ISA +15K, FL0, Mach 0.25). The aerodynamic design for some components will also be carried out at these conditions. T4 at this point is kept constant in all but a single case. T41 will vary due to off-design performance effects. The thrust produced at this point is used to deduce the ISA SLS thrust using flat take-off rating rules.
Take-off T4	K	F	F	F	F	F	F	F	F	F	F	F	F	F	F	V	F	F	F	F	
Take-off T41	K	V	V	V	V	V	V	V	V	V	V	V	F	F	V	F	F	V	V	V	
Take-off IC effectiveness	%	F	F	F	F	F	F	F	F	F	F	F	F	F	F	F	F	F	F	F	
ICAO take-off thrust	kN	V	V	V	V	V	V	V	V	V	V	V	F	F	V	F	F	V	V	V	The ICAO TO point (ISA SLS) together with the 85%F_N, 30%F_N, and 7%F_N ICAO points are used for determining the LTO NOx. Firstly, the thrust is determined at end of runway hot day TO (kink point) for a given T4, and then flat take-off rating is assumed in order to deduce the thrust at ISA SLS.
ICAO take-off T4	K	V	V	V	V	V	V	V	V	V	V	V	F	F	V	F	F	V	V	V	
ICAO take-off T41	K	V	V	V	V	V	V	V	V	V	V	V	F	F	V	F	F	V	V	V	
ICAO take-off IC effectiveness	%	F	F	F	F	F	F	F	F	F	F	F	F	F	F	F	F	F	F	F	
Mid-cruise thrust	kN	V	V	V	V	V	V	V	V	V	V	V	F	F	V	F	V	V	V	V	This is the mid-cruise specification point (ISA, FL350, Mach 0.82). The aircraft module interpolates from extended cruise performance tables, also produced by the performance module. These tables are produced using the same fixed IC effectiveness, as well as a fixed max. and min. cruise T4. As the SFC improves and/or the engine gets lighter the rubberised wing aircraft also becomes lighter. The true mid-cruise point (i.e., at the bottom of the SFC loop) will therefore vary.
Mid-cruise T4	K	F	F	F	F	F	F	F	F	F	F	F	F	F	F	F	V	F	F	F	
Mid-cruise T41	K	V	V	V	V	V	V	V	V	V	V	V	F	F	V	F	V	V	V	V	
Mid-cruise IC effectiveness	%	F	F	F	F	F	F	F	F	F	F	F	F	F	F	F	F	F	F	F	

Technology level parameter (not varied during optimisation)　　　F = Fixed　　　V = Varies

Figure 19. Main design variables for the direct drive fan conventional and intercooled core engine configuration and their effect on the values of other parameters at different off-design conditions.

Optimisation design variables (TOC unless stated otherwise)	Units	Fan inlet mass flow	Core inlet mass flow	IC LP inlet mass flow	Fan tip pressure ratio	Fan tip polytropic efficiency	IPC pressure ratio	IPC polytropic efficiency	HPC pressure ratio	HPC polytropic efficiency	HPT polytropic efficiency	IPT polytropic efficiency	LPT polytropic efficiency	IC effectiveness TOC	IC effectiveness TO	IC effectiveness cruise	IC LP pressure loss dP/P	IC HP pressure loss dP/P	T4 TOC	T4 TO	T4 cruise	T4 descent/approach/idle	HPT cooling flow % core	IPT cooling flow % core	Bypass duct pressure loss	Remarks
Mechanical/aerodynamic design	**Units**																									
LP speed	rpm	V	V	V	V	V	V	V	V	V	V	V	V	V	V	F	V	V	V	V	F	F	V	V	V	Calculated assuming a fixed fan corrected blade speed, fan face Mach number, and hub to tip ratio.
IP speed	rpm	V	V	V	V	V	V	V	V	V	V	V	V	V	V	F	V	V	V	V	F	F	V	V	V	A fixed IPC corrected blade speed is used with the tip radius being determined by components upstream of the IPC.
HP speed	rpm	V	V	V	V	V	V	V	V	V	V	V	V	V	V	F	V	V	V	V	F	F	V	V	V	A fixed HPC corrected blade speed is used with the tip radius being determined by components upstream of the HPC.
LP turbine efficiency TO	%	V	V	V	V	V	V	V	V	V	V	V	V	V	V	F	V	V	V	V	F	F	V	V	V	Varies due to off-design effects. Constant at TOC in all cases but one.
Number of IPC stages	-	V	V	V	V	V	V	V	V	V	V	V	V	V	V	F	V	V	V	V	F	F	V	V	V	Allowed to vary.
Number of HPC stages	-	V	V	V	V	V	V	V	V	V	V	V	V	V	V	F	V	V	V	V	F	F	V	V	V	Allowed to vary.
Number of HPT stages	-	F	F	F	F	F	F	F	F	F	F	F	F	F	F	F	F	F	F	F	F	F	F	F	F	Fixed.
Number of IPT stages	-	F	F	F	F	F	F	F	F	F	F	F	F	F	F	F	F	F	F	F	F	F	F	F	F	Fixed.
Number of LPT stages	-	V	V	V	V	V	V	V	V	V	V	V	V	V	V	F	V	V	V	V	F	F	V	V	V	Allowed to vary.
Objective functions	**Units**																									
Mid cruise SFC	g/(kN*s)	V	V	V	V	V	V	V	V	V	V	V	V	V	F	V	V	V	V	F	V	F	V	V	V	This is the mid-cruise specification point.
Business case block fuel	kg	V	V	V	V	V	V	V	V	V	V	V	V	V	V	V	V	V	V	V	V	V	V	V	V	
Engine weight	kg	V	V	V	V	V	V	V	V	V	V	V	V	V	V	V	V	V	V	V	V	V	V	V	V	
Technology level parameter (not varied during optimisation)		F = Fixed						V = Varies																		

Figure 20. Main design variables for the direct drive fan conventional and intercooled core engine configuration and their effect on the values of mechanical design parameters and objective functions.

Author details

Konstantinos G. Kyprianidis[1],
Andrew M. Rolt[1] and Vishal Sethi[2]

1 Rolls-Royce plc, UK
2 Cranfield University, UK

References

[1] Borradaile, J. [1988]. Towards the optimum ducted UHBR engine, *Proceedings of AIAA/SAE/ASME/ASEE 24th Joint Propulsion Conference*, AIAA-89-2954, Boston, Massachusetts, USA.

[2] Canière, H., Willcokx, A., Dick, E. & De Paepe, M. [2006]. Raising cycle efficiency by intercooling in air-cooled gas turbines, *Applied Thermal Engineering* 26(16): 1780–1787.

[3] da Cunha Alves, M., de Franca Mendes Carneiro, H., Barbosa, J., Travieso, L., Pilidis, P. & Ramsden, K. [2001]. An insight on intercooling and reheat gas turbine cycles, *Proceedings of the Institution of Mechanical Engineers, Part A: Journal of Power and Energy* 215(2): 163–171.

[4] Kyprianidis, K. [2010]. *Multi-disciplinary Conceptual Design of Future Jet Engine Systems*, PhD thesis, Cranfield University, Cranfield, Bedfordshire, United Kingdom.

[5] Kyprianidis, K. [2011]. Future Aero Engine Designs: An Evolving Vision, *in* E. Benini (ed.), *Advances in Gas Turbine Technology*, InTech, chapter 1. doi:10.1115/1.4001982.

[6] Kyprianidis, K., Colmenares Quintero, R., Pascovici, D., Ogaji, S., Pilidis, P. & Kalfas, A. [2008]. EVA - A Tool for EnVironmental Assessment of Novel Propulsion Cycles, *ASME TURBO EXPO 2008 Proceedings, GT2008-50602*, Berlin, Germany.

[7] Kyprianidis, K., Grönstedt, T., Ogaji, S., Pilidis, P. & Singh, R. [2011]. Assessment of Future Aero-engine Designs with Intercooled and Intercooled Recuperated Cores, *ASME Journal of Engineering for Gas Turbines and Power* 133(1). doi:10.1115/1.4001982.

[8] Lefebvre, A. [1999]. *Gas Turbine Combustion*, 2nd edn, Taylor & Francis, PA, USA.

[9] Lundbladh, A. & Sjunnesson, A. [2003]. Heat Exchanger Weight and Efficiency Impact on Jet Engine Transport Applications, *ISABE 2003 Proceedings, ISABE-2003-1122*, Cleveland, USA.

[10] NEW Aero engine Core concepts [2011]. http://www.newac.eu.

[11] Papadopoulos, T. & Pilidis, P. [2000]. Introduction of Intercooling in a High Bypass Jet Engine, *ASME TURBO EXPO 2000 Proceedings, 2000-GT-150*, Munich, Germany.

[12] Rolt, A. & Baker, N. [2009]. Intercooled Turbofan Engine Design and Technology Research in the EU Framework 6 NEWAC Programme, *ISABE 2009 Proceedings, ISABE-2009-1278*, Montreal, Canada.

[13] Rolt, A. & Kyprianidis, K. [2010]. Assessment of New Aero Engine Core Concepts and Technologies in the EU Framework 6 NEWAC Programme, *ICAS 2010 Congress Proceedings, Paper No. 408*, Nice, France.

[14] Walsh, P. & Fletcher, P. [1998]. *Gas Turbine Performance*, 1st edn, Blackwell Science, United Kingdom.

[15] Xu, L. & Grönstedt, T. [2010]. Design and Analysis of an Intercooled Turbofan Engine, *ASME Journal of Engineering for Gas Turbines and Power* 132(11). doi:10.1115/1.4000857.

Chapter 4

Gas Turbine Cogeneration Groups Flexibility to Classical and Alternative Gaseous Fuels Combustion

Ene Barbu, Romulus Petcu, Valeriu Vilag,
Valentin Silivestru, Tudor Prisecaru, Jeni Popescu,
Cleopatra Cuciumita and Sorin Tomescu

Additional information is available at the end of the chapter

1. Introduction

The gas turbine installations represent one of the most dynamic fields related to the applicability area and total installed power. The gas turbines have been developed particularly as aviation engines but they find their applicability in many areas, one of which being simultaneously obtaining electric and thermal energy in gas turbine cogeneration plants. The gas turbine cogeneration plants may be classified based on the constructive technology of the gas turbine in [1]: aeroderivative gas turbines plants (up to 10 MW); industrial gas turbines plants, specifically designed for obtaining energy (from 10 up to hundreds MW). An aviation gas turbine with expired flying resource is still functional due to the fact that the flight time is limited as a consequence of the specific safety normatives requirements. Therefore the aeroderivative gas turbine is defined as a gas turbine, derived from an aviation gas turbine, dedicated to ground applications. According to the initial destination, these gas turbines have been designed for maximum efficiency considering the limited fuel quantity available for an aircraft flying large distances. The basic idea in developing the aeroderivative gas turbine has been to transfer all the scientific and technologic knowledge ensuring a high degree of energy utilization (design concepts, materials, technologies, etc.) from aviation to ground [2]. Therefore the obtained gas turbines are lighter, with smaller size, increased reliability, reduced maintenance costs and high efficiency. The remaining resource for ground applications is proportional with the flight resource, being able to reach up to 30,000 hours considering the lower operating regimes. From the point of view of the actual application, the free power turbine groups are the most recommended [3]. Unlike the aeroderivative turbine power units, the industrial power units are built by the original producer with the necessary changes for actual industrial application. The development of aeroderivative and industrial gas turbines has been affected by the progress of the aviation gas turbines in military and civilian fields. Many aeroderivative gas turbines ensure compression rates of 30:1 [4]. The industrial gas turbines are cumbersome but they are more adaptable for long running and allow longer periods between maintenance controls. The base fuel for gas turbine cogeneration groups is the natural gas (with a possible liquid fuel as alternative) but the diversification of the gas turbines users and the increase in fuels price has pushed the large producers to consider alternative solutions. Nowadays the most utilized fuels in gaseous turbines are the liquid and gas ones (classic and alternative). The high temperature of the exhausted gases, approximately 590 °C on some gas turbines, allows the valorization of the heat resulted in a heat recovery steam generator. Due to the fact that the oxygen concentration in the exhausted gases is 11-16% (volume), a supplementary fuel burning may be applied (afterburning) in order to increase the steam flow rate, compared to the case of the heat recovery steam generator [5]. The afterburning leads to an increase in flexibility and global efficiency of the cogeneration group, allowing the possibility to burn a large variety of fuels, both classic and alternative. Nitrogen oxides usually represent the maine source of emissions from gas turbines. The NO_x emissions produced by the afterburning installation of the cogeneration group are different according to the system, but they are usually small and in some cases the installation even contributes to their reduction [6]. The usual methods for NO_x emissions reduction, water or steam injection for flame temperature decrease, affect the gas turbine performances, particularly to high operating regimes, leading to CO emissions increase. It must be noted that the load of the gas turbine also affects the emissions, the gas turbine being designed to operate at high loads. The general theme of the chapter is given by the technological aspects that must be considered when aiming to design a gas turbine cogeneration plant flexible from the points of view of the utilized fuel and the qualitative and quantitative results concerning some classic and alternative gas fuels. Based on the specific literature in the field and the experience of National Research and Development Institute for Gas Turbines COMOTI Bucharest, there are approached theoretic and experimental researches concerning the utilization of natural gas, as classic fuel, and respectively dimethylether (DME), biogas (landfill gas) and syngas, as alternative fuels, in gas turbine cogeneration groups, the interference between flexibility and emissions. It is particularly analysed the issue of reutilization of aviation gas turbines in industrial purposes by their conversion from liquid fuel to gas fuels operation. There is further presented the actual method of conversion for an aviation gas turbine in order to be used in cogeneration groups.

2. The aeroderivative gas turbine – A solution for gas turbine cogenerative groups flexibility on gas fuels

The flexibility of the gas turbine cogeneration plants implies reaching an important number of requirements: operating on classic and alternative fuels; capability of fast start; capability to pass easily from full load to partial loads and back; maintaining the efficiency at full load and partial loads; maintaining the emission to a low level even when operating on partial loads. Internationally, many companies with top performance in aviation gas turbines are involved in aeroderivative programs in response to market demands for energy producing installations. The best known among these are: Rolls-Royce, Pratt & Whitney, General Electric, Motor Sich, Turbomeca, MTU, etc. Rolls-Royce has developed the RB 211-H63 gas turbine starting from the aviation RB 211 which, through novel constructive and technologic transformations has been pushed to efficiency up to 41.5%. A 38 MW version will be available in 2013 with the possibility of upgrade to 50 MW in future years [7]. Many gas turbine producers aim to reach the full load in ten minutes from the start. A Japanese project of Mitsubishi Heavy Industries Ltd. (MHI) aims to manufacture a gas turbine operating at 1700 °C inlet temperature and 62 % efficiency. Pratt & Whitney, starting from the PW 100 turboprop, have developed the ST aeroderivative gas turbine family (ST 18, ST 40). The researches conducted at National Research and Development Institute for Gas Turbines COMOTI Bucharest have allowed obtaining aeroderivative gas turbines in the 20 – 2,000 kW range, through valorisation of the aviation gas turbines with exhausted flight resource, obsolete or damaged. Therefore the AI 20 GM (figure 1, right) aeroderivative turboshaft, operating on natural gas, is based on the AI 20 turboprop (figure 1, left). The AI 20 GM is used in power groups driving the backup compressors in the natural gas pumping stations on the main line at SC TRANSGAZ SA. The aeroderivative GTC 1000 (figure 2, right), based on TURMO IV C (figure 2, left), operating on natural gas, is used in a power group driving two serial centrifugal compressors for the compression of the associated drill gas, in one SC OMV PETROM SA oil exploitation, at Țicleni – Gorj. Researches have also been conducted regarding the valorisation of the landfill gas in a aeroderivative gas turbine applicable to cogeneration groups [2]. A project for a cogeneration application using the GTE 2000 aeroderivative gas turbine has been started in 2000. The result of the project is a cogeneration plant, with two independent lines, producing electric and thermal (hot water) energy, located in the municipality of Botosani, with SC TERMICA SA as beneficiary (figure 3, left). The experience acquired from the GTE 2000 cogeneration plant has been used in a new project for a medium power aeroderivative gas turbine cogeneration plant, the application using the ST 18 A aeroderivative gas turbine, manufactured by Pratt & Whitney. The ST 18 A aeroderivative gas turbine has been derived from the aviation PW 100 through redesigning a series of components of which are distinguished the combustion chamber, the case and the intake. Furthermore, the ST 18 A has been designed and manufactured to operate with water injection in the combustion chamber (duplex burners), method that ensures the reduction of NO_x emissions. The application consists in a cogeneration plant, with two independent cogeneration lines, producing electric and thermal (superheated steam used in the oil extraction technologic process) energy. The beneficiary of the application is SC OMV PETROM SA, Suplacu de Barcău, Bihor County (figure 3, right) [8]. What makes the difference between aviation and aeroderivative gas turbines are operating conditions and reliability. Thus, aviation gas turbines have over the period of their useful life so many ordered starts and stops (associated with aircraft flight), short operation between starting and stopping (of hours), short periods between revisions (after each stop) and overhauls (after more than 1,000 hours of operation), the lifespan of about 12,000 cumulative hours of operation. Aeroderivative gas turbines can operate up to 8,000 hours continuously without ordered stop, overhauls are made at intervals up to 30,000 cumulative operating hours and, for some brands, the cumulative operating ranges may be even higher.

Figure 1. AI 20 turboprop (left) and AI 20 GM gas turbine (right) [2]

Figure 2. TURMO IV C turboshaft (left) and GTC 1000 gas turbine (right) [2]

Figure 3. GTE 2000 – Botoşani (left) and 2xST 18 – Suplacu de Barcau (right) plants

2.1. Classic and alternative fuels for gas turbine cogeneration groups

The performances of the gas turbine cogeneration groups (efficiency and emissions) depend in high degree of the type and physical and chemical properties of the used fuels. Depending on the lower heating value (LHV), in relation to natural gas (LHV=30-45 MJ/Nm^3), typical gas fuels can be classified as [9]: high heating value (LHV=45-190 MJ/Nm³; butane, propane, refinery off-gas), medium heating value (LHV=11.2-30 MJ/Nm³; weak natural gas, landfill gas, coke oven gas), low heating value (LHV<11.2 MJ/Nm³; BFG - Blast Furnace Gas, refinery gas, petrochemical gas, fuels resulted through gasification etc).

2.1.1. General requirements regarding the utilization of fuels in gas turbines

For the gas turbines used in cogeneration groups, for economic reasons, the most used fuels are heavy oil and waste products from various manufacturing or chemistry processes [3]. Using liquid fuels imposes: ensuring combustion without incandescent particles and deposits on the firing tube and the turbine; decreasing the corrosive action of the burned gases caused by the aggressive compounds (sulphur, lead, sodium, vanadium, etc.); solving the pumping and atomization issues (filtration, heating, etc.). A series of fuels must be well purified or filtrated for eliminating water, solid particles or some remiss substances. Heavy liquid fuels must be heated to a convenient temperature to allow their proper pumping and spraying. Coke number and tar number are of particular interest for burning in gas turbines. Coke number (carbon residue) represents the residue left by an oil product (fuel oil, diesel, etc.) when burned in special conditions (closed space, restricted air access, etc.), expressed in mass percent. Tar number indicates the presence of resins, aromatic hydrocarbons, etc. but it must be considered for information only. In order to define the combustion behaviour of a heavy liquid fuel (like oil) it would be indicated to consider as a criterion the product of the coke number and tar number [10]. In terms of reusing aviation gas turbines in industrial purposes, the possibilities of using liquid fuels are decreasing. For each application, the requirements of the beneficiary must be analysed related to the characteristics of the fuels affecting the combustion (density, molecular weight, evaporation limit, flammability temperature, volatility, viscosity, surface tension, latent heat of vaporization, calorific value, the tendency for soot, etc.). In terms of using gas fuels, the problem is less challenging due to their thermal stability, high heating value, lack of soot and tar. However, in order to ensure the pressure level required by the gas turbine, afterburning, etc., the elimination of water and different impurities, a control – measurement station must be provided for the gas fuels to be used (natural gas at 2xST 18 plant – figure 4). Some alternative gas fuels (resulted

Figure 4. Control – measurement station for natural gas at 2xST 18 – Suplacu de Barcau plant (left) with booster (right) 1 – cogeneration power plant; 2 – control – measurement station; 3 – booster

through gasification and biomass pyrolysis), biogas, residual gases from industrial processes (rich in hydrogen) can play an important role in the operation of the gas turbine cogeneration groups, but they must reach some requirements regarding the calorific value and the composition [11]. Therefore there is necessary to eliminate the impurities, tar, to limit the sulphur and its compounds to 1 mg/Nm^3, respectively the alkaline metal compounds to 0.1 mg/Nm^3 [12].

2.1.2. Alternative fuels – Characteristics and consequences regarding their use in gas turbine cogeneration groups

The biogas produced through anaerobic fermentation is cheap and constitutes a renewable energy source producing, from burning, neutral carbon dioxide (CO_2) and offering the possibility of treatment and recycling for residues and secondary agricultural products, various biowaste, organic waste water from industry, sewage and sewage sludge. The properties and the composition of biogas are different depending on the raw material used, processing system, temperature, etc. The comparative compositions of natural gas and biogas are given in table 1 [13]. For both fuels the main component (giving the energetic value) is the methane (CH_4), the significant differences being given by the high content of CO_2 and H_2S (hydrogen sulphide) in biogas. Technically, the main difference is given by the Wobbe index for natural gas (see chapter 2.2), two times higher than the index for biogas. This leads to a limited possibility of replacing the natural gas with biogas because only gases with similar Wobbe index can substitute each other. The improvement of the biogas can be achieved by replacing CO_2 with CH_4 so as to approach the characteristics of natural gas. Furthermore, the water and hydrogen sulphide must be eliminated to avoid the harmful action of the resulted sulphuric acid on different components of the cogeneration group (gas turbine, afterburning installation, heat recovery steam generator, etc.). Landfill gas resulted from waste deposits represents a cheap energy source, with a composition similar to the biogas resulted from anaerobic fermentation (45-60 % methane, 40-55 % carbon dioxide) [2]. When it comes to using biogas in gas turbine cogeneration groups or introducing it in the natural gas network, special treatment is required (condensate separation, drying, adsorption of volatile substances, etc.). Dimethylether (DME, CH_3-O-CH_3) is a clean alternative fuel which can be produced from fossil fuels, namely coal or vegetal biomass gasification. It can be transported and stored similar to liquefied petroleum gas (LPG), its physical and chemical characteristics, related to natural gas in Ardeal (99.8 % CH_4 and 0.2 % CO_2), being given in table 2 [14]. The flame produced by burning the dimethylether is very similar to the flame produced by the natural gas (figure 5), which makes it suitable to be used as fuel in transportation, cogeneration groups, etc.

Through biomass of coal gasification (with oxidant agents such as oxygen, air, steam, etc.) it can obtain synthesis gas (syngas) with main components hydrogen (H_2) and carbon monoxide (CO). The syngas can be used to obtain methanol, hydrogen, methane, etc. or can be used as fuel in gas turbine cogeneration groups. Since leaving the gas-producing installation the gas contains ash particles and various compounds of chlorine, fluorine, alkali metals, etc., which must be removed to protect the cogeneration line. Through gasification of different biomass categories and utilization of different gasification technologies, the composition of the resulted gas and the lower heating value (LHV) can vary according to tables 3 and 4 [12, 15]. Tables 1 and 3 show that the lower heating values for biogas and syngas are lower than for the natural gas, requiring, in their application in cogeneration groups, higher mass flow rates with minimum pressure losses. Therefore, the injection nozzles of the gas turbine and the burners of the afterburning installation must be designed for velocities allowing a homogenous mixture between fuel and oxid, as well as low pressure losses. The syngas contains high quantities of hydrogen which affect the combustion in gas turbine cogeneration groups in terms of flame stability, combustion efficiency, etc. Using hydrogen as fuel and introducing a component with dilution role (steam, nitrogen, etc.) the operation of the gas turbine is affected [16].

No.	Name	Natural gas	Biogas
1	CH_4 [vol %]	91.0	55-70
2	C_nH_{2n} [vol %]	8.09	0
3	CO_2 [vol %]	0.61	30-45
4	N_2 [vol %]	0.3	0 - 2
5	Lower heating value [MJ/Nm³]	39.2	23.3
6	Density [kg/Nm³]	0.809	1.16

Table 1. Composition, physical and chemical proprieties for natural gas and biogas [13]

No.	Name	Natural gas (Ardeal)	Dimethylether
1	Theoretical combustion temperature [ºC]	1,900	2,000
2	Autoignition temperature [ºC]	650-750	350
3	Lower heating value [MJ/Nm³]	35.772	59.230
4	Explosion limit [% gas in air]	5 - 15.4	3 - 18.6
5	Density [kg/Nm³]	0.716	2.052

Table 2. Physical and chemical characteristics for natural gas (Ardeal) and dimethylether [14]

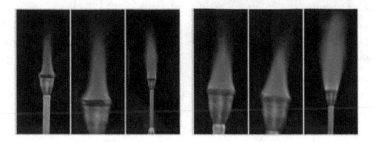

Figure 5. Flame of Bunsen burner, with grid type flame stabilizer, on natural gas (left) and dimethylether (right) [14]

	Syngas chemical composition [%]						Lower heating value
Name	**CO**	**H₂**	**CH₄**	**CₙH₂ₙ**	**CO₂**	**N₂**	**[MJ/Nm³]**
Dry oak	18.3	16.9	2.8	0.5	16.0	-	5.422
Dry beech	19.4	17.5	2.6	0.6	15.0	49.3	5.526
Dry fir	15.1	19.1	1.6	0.9	15.8	57.1	4.053
Wood coals	31.2	6.3	2.9	-	2.5	57.1	5.702

Table 3. Chemical composition of syngas and lower heating values resulted from biomass gasification [15]

CO [%]	H2 [%]	CH4 [%]	N2 [%]	H₂O [%]	CO2 [%]	LHV [MJ/Nm³]	Observations
16	6	4	56	18	-	4.1	Air gasification
16	6	4	56	15	3	4.1	Air gasification
40	13	15	3	-	29	11.826	Oxygen gasification

Table 4. Chemical composition of syngas and lower heating values resulted from different methods of gasification [12]

Solving the fuels interchangeability issue for gas turbine cogeneration groups, by developing high level combustion technologies for alternative fuels, particularly hydrogen, will have a major impact on system efficiency and environment.

2.2. Fuels interchangeability and validation criteria

Interchangeability in gas turbine cogeneration groups represents the capability to replace a gas fuel with another without affecting the application or the installation burning the gas fuel. The used gas fuels consist in mixtures of combustible gases (methane and other light hydrocarbons, hydrogen, carbon monoxide) and inert gas (mostly nitrogen, carbon dioxide, water vapor). Depending on the combustible gases ratio (usually methane), the gas fuels can have high or low heating value. Density and temperature of the used fuel, as well as the environmental temperature, can affect the performances and lifespan of the equipments in the cogeneration group. According to these influence factors, the most important parameter for characterizing the interchangeability is the Wobbe index (named after engineer and mathematician John Wobbe), defined as ratio between the lower heating value (LHV) and the sqare root of density of the fuel, relative to air density (d_{rel}):

$$Wo = LHV / (d_{rel})^{0.5} \tag{1}$$

$$d_{rel} = \rho_{comb} / \rho_{air} \tag{2}$$

Therefore, two gas fuels, with different chemical compositions but the same Wobbe index, are interchangeable and the heat delivered to the equipment is equivalent for the same fuel pressure. Table 5 gives the values of Wobbe index for several gas fuels. In order to consider the temperature of the fuel, the Wobbe index can be corrected with the temperature. According to [17], two fuels are interchageable if they respect:

$$\frac{\Delta p_2}{\Delta p_1} = \left(\frac{Wo_1}{Wo_2}\right)^2 \left(\frac{A_1}{A_2}\right)^2 \tag{3}$$

where Δp_1 and Δp_2 represent the overpressure of fuel 1, respectivelly 2, Wo_1 and Wo_2 – Wobbe indexes of fuel 1, respectively 2, A_1 and A_2 – injection nozzle area for the two fuels.

No.	Gas name	Wobbe index [(MJ/Nm³]
1	Natural gas	48.554
2	Liquefied petroleum gas	79.993
3	Methane	47.947
4	Ethane	62.513
5	Propane	74.584
6	Carbon monoxide	12.812
7	Biogas	27.3
8	Dimethylether	47.422
9	Hydrogen	38.3

Table 5. Wobbe index for various gases [2, 13, 14]

Therefore, the validation criteria for replacing a fuel with an equivalent one are given by: autoignition temperature, flame temperature (with higher influence on NO_x formation), flame velocity, flashback, efficiency, NO_x and CO emissions, flue gas dew point, etc. Autoignition temperature of gas fuel in mixture with air is the temperature on which the instantaneous and explosive autoignition occurs, without the existence of an incandescent source of ignition. The turbulent flame is generally less stable than the laminar flame, the instability in flame front break-up field being emphasized by the increase in tube diameter. Free swirl turbulent flames are more prone to flame front break-up than the laminar ones due to the higer periferal jet velocity. For turbulence angles greater than 30°, the stability area is achieved on the contour of the burner only for rich mixtures [18]. In areas with poor mixture, due to the decrease in velocity, the backflow can occur without flame attachment on the burner edge. The velocity distribution in the swirl flow determines the stabilisation of the flame as a central suspended one. Components with rapid burning, such as hydrogen, accelerate the flame velocity with a tendency to backflow or extinguishment. The backflow tendency of the flames is proportional with the ignition velocity of the fuel gas, a high velocity leading to a high effect. It is also dependent of the primary air proportion and the components with reduced burning velocity can lead to flame front break-up. In order to consider these factors, an empiric relation has been established for the flame front break-up index at interchangeability I_{ret} [19]:

$$I_{ret} = \frac{k_i f_i}{k_b f_b} \left(\frac{LHV_i}{LHV_b} \right)^{0.5} \tag{4}$$

where: k – constant concerning the flame front break-up limit; f – factor concerning primary air; LHV – lower heating value; b and i – indexes regarding the control fuel, respectively the replacement fuel. A particular issue is raised by the fuels with reduced heating value. Therefore, the landfill gas contains over 40 % CO_2, requiring a suitable fuel feeding in order to achieve combustion. The fuels with reduced heating value have a small range of flammability requiring, at partial loads or transient operating regimes, the utilization of a supplementary fuel (such as propane). The mass flow rates necessary for gas turbine operation on reduced heating value gas fuels are high (neglecting the water or steam injection in the gas turbine) compared with the operation on natural gas, fact that modifies the compressor's operating characteristic [20]. From biomass gasification with air, it is obtained syngas with LHV of 4-6 MJ/Nm^3, and from the gasification with steam or oxygen (see table 4) LHV of 9-13MJ/Nm^3. An alternative for increasing lower heating value is the mixing with natural gas. Therefore, if the landfill gas has a LHV of 17-20 MJ/Nm^3, an equivalent lower heating value can be obtained by mixing 60 % gas with reduced heating value with 40 % CH_4, with respect to the composition described in [21].

2.3. Converting the aviation gas turbines from liquid to gas fuels operation

The complexity of thermo-gas-dynamic processes defining the gas turbine operation in a cogeneration group require theoretical and experimental research activities on gas turbines in order to accomplish the conversion from liquid to gas fuels operation. For the gas turbines on market, in exploitation, the exploitation and maintenance technical specifications are generally known, being provided by the producer. When the object of the research is an existing gas turbine lacking the technical documentation which completely define the contructive solution, the issue must be approached through activities of experimentation, measurements, CAD 3D modelling, numerical simulation in CFD environment, constructive modifications and renewed experimentation in order to validate the constructive solutions, permanently aiming the performances correlated with the maximum effectiveness (thrust, power), minimum specific fuel consumption, maximum efficiency, versatility on fuel conversion, maximum availability, minimum operation and maintenance costs.

2.3.1. General criteria – Researches concerning the modifications on a gas turbine for gas fuel operation

The basic procedure for an aeroderivative gas turbine is to keep the rotor assembly, compressor – turbine, which is the „heart" of the gas turbine, form the aviation gas turbine and to redesign the combustion chamber in order to operate on a different fuel than the kerosene. Therefore, for the basic gas turbine in the turboshaft category, at least the combustion chamber must be designed for gas fuels operation. The shaft of the power turbine is mechanicaly connected to a driven load, mechanical work consumer, depending on the application involving the aero-derivative gas turbine (electric generator, compressor, pump, etc.). The command and automatic control system of the aero-derivative gas turbine are designed depending on the application. The bearings can be redesigned, achieving a conversion from rolling bearings to slide bearings. For the basic turboprop (destined for propeller aircrafts), at least the combustion chamber and the reducing gear box and/or the gas generator's turbine must be redesigned, depending if the turboprop does or does not include free turbine. Usually, only the gas generator is used, eliminating the gear box. The issues concerning the automatic control system and the bearings are identical to those of the turboshaft. For the basic turbojet (simple flow jet predominantely for military aircrafts) the redesigning of the combustion chamber and the designing of a power turbine gas-dynamicaly connected to the gas generator are necessary [2]. The issues concerning the automatic control system and the bearings are also identical to those of the turboshaft. Regarding the combustion chamber, is desired to constructively alterate it as little as possible, maybe only in terms of injection system. Due to the fact that the rest of the parameters characterizing the operating process remain unchanged, those regarding zero velocity and ground conditions of the basic gas turbine, the operation of the combustion chamber can be considered as in terms of gas-dynamic similarity. A first problem that must be studied when replacing the fuel is maintaining the combustion efficiency. A second one concerns the maintaining of constructive-functional temperature distribution (on the walls of the firing tube, in the outlet area of the combustion chamber and inlet area of the turbine). On the background of the assembly gas-dyanmic characteristics, the unevenness of the temperatures field on the outlet of the combustion chamber (temperature map) is determined by the geometric characteristics of the dilution area (diameter, length, number and area of holes, etc.) and the characteristics of fuel feeding in the primary area (atomization, jet angles, fuel specifications, etc.). The global temperature map is defined by equation (5) and the radial unevenness for the rotor bladed area is given by equation (6) [3]:

$$\theta_m = \left(T^*_{max} - T^*_3\right) / \left(T^*_3 - T^*_2\right) \qquad (5)$$

Radial unevenness for the rotor bladed area express the manner of operation on the turbine blades:

$$\theta_r = \left(T^*_{maxr} - T^*_3\right) / \left(T^*_3 - T^*_2\right) \qquad (6)$$

In equations (5) and (6) the significance of symbols is: T^*_{max}- maximum temperature peak; T^*_3- average temperature in the outlet section of the combustion chamber; T^*_2- average temperature in the outlet section of the compressor; T^*_{maxr}- maximum average radial temperature, circumferential arithmetic mean on the entire section. Normal values for θ_m, depending on the gas turbine, are in the 20-25 % range, with reported values of 35 %. In direct connection with the temperature map on the walls of the firing tube, the equivalent stress of the material must be considered when replacing a fuel with another. In the case of the AI 24 gas turbine modification for operation on gas fuels in the cogeneration group, a difference of 15 % has been reported in the temperature map, considering the flattening of the temperature peaks when passing through the turbine [22]. The adopted solution has been the generalization of the results obtained by National Research and Development Institute for Gas Turbines COMOTI Bucharest for the AI 20 GM (figure 1) and MK 701 gas generators. In order to achieve the AI 20 GM gas turbine on natural gas (derived from AI 20 on liquid fuel) the adopted constructive solution has been the modification of the injection system, without altering the firing tube (figure 6). The researches for this transformation have been based on test bench experiments with liquid fuel (in low pressure similitude conditions). In order to reach the functional optimum on natural gas, several injection nozzles have been designed and experimented, according to table 6 [3].

Nozzle no.	10 Ø3 holes at a 2α angle	Diameter of central hole Ø [mm]
1	90^0	3
2	70^0	without central hole
3	80^0	without central hole
4	70^0	3
5	80^0	3
6	100^0	without central hole

Table 6. Configuration of the experimental injection nozzles (see figure 6), for AI 20 GM on natural gas [3]

Only nozzles with 10 holes of the same diameter have been experimented in order to ensure velocity, penetration and safety in operation. The central hole afects the stability of the combustion process, increases the flame radiation and the temperature on the walls of the firing tube. The tie criterion for various injection nozzles for natural gas has been the temperature of the blade on hub. It has been noted that nozzle no. 3 leads to low frequency vibrations in a large range of operating regimes, functionally inadmissible. When operating on natural gas, the combustion efficiency increases with the operating regime, the process being unaffected by the vaporization, but only by the mixing. Following the experimentation, nozzle no. 2 has been selected (with 10 Ø3 holes at $2\alpha=70^0$, without central hole). For all experimentation regimes, the circumferential temperature map values did not pass 18 %. The same manner of minimum configuration modifications has been applied for the rest of the gas turbines transformed for operating on natural gas (TURMO, MK 701, etc.). Therefore, the firing tube and the combustion chamber case have been kept unmodified for all gas turbines, only redesigning the injection system. Satisfying results have been obtained for the experimentation of TURMO: good stability, but in a more limited range compared with other gas turbines (due to the dependency on the mixing process); temperature map values of 22 % (for the aimed 20 %). For MK 701, the values on the temperature map have reached max. 20 %. A particular problem is considered when the aim is the integration of the gas turbine, modified for operating on natural gas, with an existing boiler. The heat recovery steam generator can be derived form an energy steam boiler, a technological steam boiler or a hot (warm) water boiler. The integration analysis for an aeroderivative gas turbine with a hot water boiler shows that the temperature of the burned gases on the stack must be in the usual value range and the pressure loss at the passing through the modified boiler (in the cogeneration group) must be lower than the pressure loss on the initial boiler [23]. The modifications necessary for operating the gas turbine on gas fuels with reduced lower heating value, compared with the operation on natural gas, are slightly more complex. Therefore, Mitsubishi Heavy Industries Ltd., with extensive experience in manufacturing gas turbines on BFG (Blast Furnace Gas), considerd the heating value of the gas fuel as the key factor in the modifications scheduled for the gas turbine [24]. Depending on the actual application, more modifications can be operated on the gas turbine, compared with the ones in table 7.

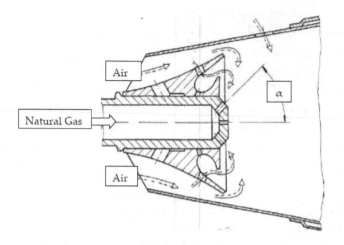

Figure 6. Modification of the injection system for Al 20 GM gas turbine [3]

Lower heating value [MJ/Nm³]	20.95-41.9 (High)	35.61 (Natural gas)	8.38-29.33 (Medium)	2.51-8.38 (Low)
Air compressor	Standard	Standard	Standard	Modification
Combustor	Standard (Minor mod.)	Standard	Standard (Minor mod.)	Modification
Turbine	Standard	Standard	Standard	Standard
Fuel system	Standard (Minor mod.)	Standard	Standard (Minor mod.)	Modification

Table 7. Necessary modifications for a gas turbine, depending on lower heating value of the fuel [24]

2.4. Converting a gas turbine from liquid to gas fuel operation for landfill gas valorisation

Converting the gas turbine from liquid fuel to gas fuel operation in order to achieve the valorisation of the landfill gas has known two main steps, respecting the principles in chapter 2.3: converting the TV2-117A gas turbine from operating on liquid fuel (kerosene) to gas fuel (natural gas), resulting the TA2 gas turbine; converting the TA2 from operating on natural gas to operating on landfill gas, resulting TA2 bio. In order to achieve these results, numerous numerical simulations in CFD environment and tests have been used for validating the adopted solutions.

2.4.1. Numerical simulation, experimental activity, methods and equipments

Numerical simulation on the TV2-117A gas turbine (figure 7, left) on kerosene has been made in order to obtain a reference model for the gas turbine conversion on gas fuels, particularly landfill gas. An eighth of the geometric model, corresponding to one injection nozzle, has been used in simulations considering the combustion chamber simetry. The boundary conditions have been provided by the producer in the technical specifications for three operating regimes: take-off, nominal and cruise (with the corresponding temperatures of 1123, 1063 and 1023 K). For simulating the combustion process in the TA2 bio gas turbine, the used fuel has been a synthetic landfill gas with equal volume proportions of methane (CH_4) and carbon dioxid (CO_2). The real landfill gas contains other chemical species, in small proportions, which have been considered impurities and have not been taken into account. The numerical simulations have been made on the TA2 with modified injection system, particularly on the injection nozzles level (figure 8). The modelling of the injection nozzles has been achieved starting from the geometry of the natural gas nozzles. Only the injector's outer body have been kept from the liquid operating gas turbine, eliminating all elements related to the atomization system of the liquid fuel. Related to the initial configuration of the injector, only the diameter of the secondary channel and the configuration of the connection with the injection nozzle have been kept unmodified.

Figure 7. TV2-117A gas turbine (left) with detailed combustion chamber area (right)

The numerical simulations for the modified injector (figure 8) have taken into consideration the variation of the injection pressure (7.65 - 8.5 bar), of the injection angle β (70 - 85⁰) and the position related to the injector's body L (1 - 5 mm). Following the numerical simulations, the optimum configuration has been selected and the eight injectors have been manufactured along with the injection ramp (figure 10, right), consisting in a circular pipe connected to each injector. The configurations of the injectors for liquid fuel and landfill gas are given in figure 9. The elements eliminated from the initial configuration are the following: the liquid fuel feeding system; the liquid fuel automatic control system; the command system for the actuators controlling the guide vanes and the first three statoric stages of the compressor; the deicing system. The experimentation of TA2 bio has been made in the experimental facility of National Research and Development Institute for Gas Turbines COMOTI Bucharest (figure 10) in the following configuration: TA2 bio gas turbine installed on test bench; test cell lubricating system and fuel feeding system for the gas turbine; exhaust system for the burned gases; monitoring system for acquiring functional parameters. In figure 10 (right) is a pipe ramp ring, yellow color, for gas fuel supply.

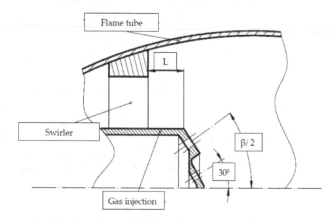

Figure 8. Injection nozzle configuration for landfill gas [2]

Figure 9. Injectors for liquid fuel for TV2-117A (left) and landfill gas for TA2 bio (right) [2]

A series of experimentations have been made, the simulated landfill gas being obtained by mixing natural gas with carbon dioxid (provided from tanks). The measurements have been made with the equipments of the test facilities. A ramp of 17 double thermocouples located at the outlet of the combustion chamber, with measuring points at one third and two thirds of the outer firing tube circumference allow the measurement of the T_{ex} and T_{in} temperatures on two concentric rings (figure 11 right).

Figure 10. TA2 gas turbine (left) and TA2 bio gas turbine in the test cell (centre, right)

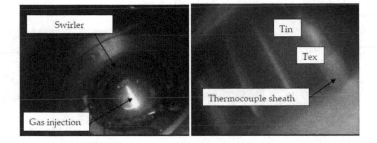

Figure 11. Boroscoping images of the gas injection nozzles – natural gas (left) and the thermocouples (right)

2.4.2. Results and discussion

The numeric simulations on kerosene [2, 5] have shown that, for the reduced operating regimes, the flame reaches in high degree the area between two adjacent injectors. Table 8 presents the numerical results for landfill gas combustion in terms of methane mass fraction, illustrating the jet shape, and burned gases temperature in the oultlet section of the combustion chamber. Analysis of data in table 8, with respect to temperature maps, aiming to obtain a compact jet in order to protect the walls of the firing tube, have helped selecting the geometric configuration of the injection nozzle: $\beta = 70^0$ and L= 3 mm, used for designing the functional model experimented on TA2 bio, for a mixture of natural gas and carbon dioxide. The experiments have been developed in several series, figure 12 presenting one of the models of variation for the components of the synthetic landfill gas mixture. The experimental results have been synthetized in figures 12 and 13. Figure 13 presents the numerical and experimental results for the outlet section of the combustion chamber.

Parameters			Results	
L [mm]	β [°]	p [bar]	Fuel injection jet	Temperature on combustion chamber outlet
1	70	7.65		
3	70	7.65		
5	70	7.65		
3	80	7.65		
3	85	8.50		

Table 8. Numerical results for landfill gas combustion simulation [2]

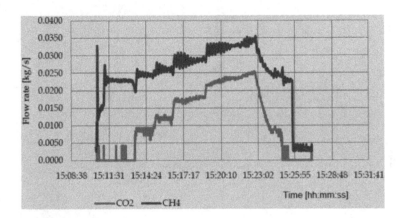

Figure 12. Variation of the mass flow rates of carbon dioxide (CO_2) and natural gas (CH_4) injected in the combustion chamber [2]

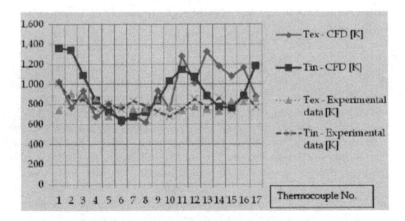

Figure 13. Comparison between the numerical and experimental temperature

The experimentations have proved a stable operation of the TA2 bio gas turbine on different operating regimes, mainly defined by the mass flow rate and the ratio between the mass flow rate of the natural gas and carbon dioxid. Figure 13, particularly the central area, shows a concordance of the numerical and experimental data, proving that modification of gas turbines operating on alternative gas fuels can be made based on numerical simulations in CFD environment. The model of a cogeneration plant for electric and thermal energy is illustrated in figure 14.

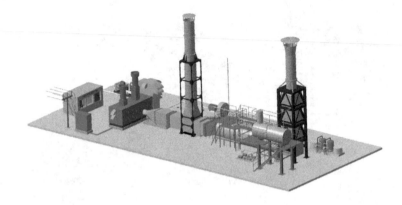

Figure 14. Model of an aeroderivative gas turbine cogeneration plant operating on natural gas and landfill gas

3. Flexibility of gas turbine cogeneration groups and emissions reduction – Future researches

Gas turbine cogeneration groups, alone or in combination with fuel cells, can play an important role in the general assembly of energy production and emissions reduction. The NO_x reduction must be regarded considering the ensurance of cogeneration group performances in a flexible manner, optimization being possible for a fuel [25]. A higher efficiency implies the optimization of the entire cogeneration plant (gas turbine, afterburning, heat recovery steam generator, etc.). The efficiency must be maintained for partial loads (even below 50 %) or for environmental conditions modification. Starting from 2002, Siemens has taken into consideration the flexibility, eliminating the high pressure barrel of the heat recovery steam generator which requires a long process to reach a certain temperature (in order to avoid the occurence of thermal tensions). Regarding the flexibility, the efficiency and the emissions reduction in gas turbine cogeneration groups, important steps have been made: reduced NO_x burners have been introduced in applications; the lifecycle has been analyzed for efficiency increase; the period between maintenence controls has been extended and the conversion from one fuel to another for multi-fuel engines has improved [7]. The factors determining the formation of pollutant agents exhausted along with the burned gases from the gas turbines are [26]: temperature and air excess coefficient in primary area; homogenization of the process in primary area; residence time of the products; "freezing" characteristic of the reaction near the firing tube, etc. For NO_x reduction, the temperature in the area of the combustion reaction and the areas of maximum temperature and the air jets distribution (stage combustion) need to be reduced. The final configuration of the combustion chamber of a gas turbine is a compromise between the NO_x level, performance and flexibility. Global reduction of the emissions leads to compromises between the emission levels of different components and the assembly characteristics of the combustion chamber (pressure losses, stability and ignition limits, etc.). New concepts must be promoted in order to solve this issue. The usual methods are represented by the water or steam injection in the combustion chamber of the gas turbine, leading to [12]: reduction of NO_x up to 25 ppm (for a 15 % O_2 volume participation in dry burned gases); increase in turbine power due to the increase in fluid mass flow rate (which can compensate the effect of increased temperature during summer); increase of flexibility of the installation in exploitation due to the possibility of load variation through steam flow rate variation. However, the high content of vapours in burned gases can lead to: acid corosion occurence (for fuels containing sulphure); increase in thermal stress on the combustion chamber; reduction of the heat recovery level, etc. Numerical simulations on TV2-117A (figure 15) for water injection in the combustion chamber (through duplex injectors, on natural gas) have shown that the water injection in truncated cone shape, at 45°, characterized by a 12 l/min mass flow rate, leads to minimum NO_x concentration in burned gases of 14 ppm. The analysis of combustion products for TA2 (see chapter 2.4), using NASA CEA program [27], has shown a decrease of the average maximum temperature. The composition of the landfill gas has been considered in equal volume proportions of methane and carbon dioxide, while the composition of the syngas has been considered that given by [19]. The calculation algorythm has started from the stoichiometric reaction of each fuel and imposing the operating regime (in terms of average maximum temperature of 1063 K for nominal regime) in order to determine the minimum quantity of air necessary for the reaction. Obtaining the equilibrium reactions has determined the calculation of the air excess coefficients for each fuel at the given regime, for dry operation. Starting from these initial values, water has been introduced in different proportions, up to 23 %. The supplementary quantity of fuel, necessary to reestablish the operating regime of the gas turbine, in terms of temperature (considering the pressure as unaffected), has been calculated in relation to the quantity of water. The general combustion reactions for each fuel, for the water injection case, for the nominal operating regime, are given by equation (7) for landfill gas and equation (8) for syngas:

$$b \cdot (CH_4 + CO_2) + 2 \cdot \lambda \cdot (O_2 + 3.76\ N_2) + a \cdot 2 \cdot \lambda \cdot H_2O \rightarrow w\ H_2O + x\ CO_2 + y\ N_2 + z\ O_2 \tag{7}$$

$$b \cdot (0.25 \cdot CO + 0.09 \cdot CO_2 + 0.12 \cdot H_2 + 0.52 \cdot N_2 + 0.02 \cdot CH4) + 0.225 \cdot \lambda \cdot (O_2 + 3.76\ N_2) + a \cdot 0.225 \cdot \lambda \cdot H_2O \rightarrow w\ H_2O + x\ CO_2 + y\ N_2 + z\ O_2 \tag{8}$$

There have been tracked the thermodynamic of the system and the concentrations of the reaction products, focusing on carbon monoxid (CO) and nitrogen oxides (NO_x). In these conditions, for the two regimes, the calculations have been made up to a injected water coefficient (noted „a") in oxidant of maximum 2, equivalent to 23 % water in oxidant. The maximum proportion of water in oxidant has been limited by the concentration of oxygen resulted from the combustion, minimum 11 %, necessary for the afterburning process. For the nominal operating regime and approximately 15 % water for landfill gas and 12.5 % for syngas, the gas turbine reaches the minimum limit of oxygen.

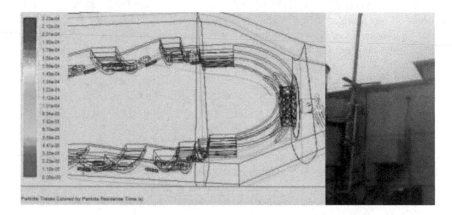

Figure 15. Numerical simulation of water injection in the combustion chamber of TA2 (left) and atomization tests with the duplex injector (right)

Figure 16 shows the variation of NO_x for the two fuels (landfill gas and syngas) for the nominal regime, depending on the injected water proportion. The results of the calculations illustrate that the use of afterburning along with the operation of the TA2 gas turbine, with water injection, for the good operation of the system, the NO_x produced by the gas turbine at 1063 K can only be reduced to 40 ppm for landfill gas and 38.5 ppm for syngas. The oxygen injected in the air can lead to nitrogen oxides reduction and combustion enhancement resulting [28]: reduction of ignition temperature; increase in flamability limit; increase in adiabatic temperature of the flame; increase in process stability and control; reduction of low heating value fuels consumption, etc. The adiabatic temperature of the flame increases with approximately 50 °C for 1 % increase in oxygen concentration. The volume of burned gases decreases with 12 % for the combustion of natural gas in 3 % oxygen enriched air [29]. Reduction of pollution through combustion in oxygen enriched environment can be used in afterburning installations (for primary or secondary air). Combustion in oxygen enriched environment can increase the efficiency and the flexibility of the cogeneration plant. When adding hydrogen to a gas fuel, there are affected the stability of the flame, the efficiency of the combustion and the emissions. Flame velocity for hydrogen combustion in air, in stoichiometric conditions, reaches 200 cm/s compared to the combustion of methane in air, for which the velocity is approximately 40 cm/s [29]. Adding hydrogen to the gas fuel of the gas turbine or afterburning installation can lead to CO and NO_x emissions reduction.

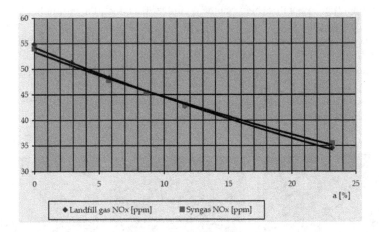

Figure 16. Variation of NO_x concentration for the two fuels, at 1063 K, depending on water proportion in oxidant (a)

3.1. Afterburning installation as interface between gas turbine and heat recovery steam generator

The burned gases flow when exiting the gas turbine is turbulent and unevenly distributed in transversal section. Therefore, backflow can occur in the transversal section of the recovery boiler. The unevenness of the flow and the variation in burned gases composition affects the operation of the afterburning. Therefore, the afterburning is influenced in terms of efficiency, emissions, flame stability, as well as corrosion of the elements subjected to the action of burned gases. For a good design of the inlet section in the recovery boiler it must be generally considered the following factors [30]: geometry and direction of the gas turbine exhaust; size of heat exchange surfaces; location of the afterburning burner; mass flow rate and average velocity of burned gases exiting the gas turbine; local velocities near the walls and on the first heat exchange surface. The gas turbine exhaust is generally not directly connected with the recovery boiler. After exiting the gas turbine (the case of 2xST 18 Cogeneration Plant at Suplacu de Barcau), the burned gases pass through a silencer, a by-pass assembly, a transom for the connection with the burner and then the afterburning chamber [8]. The gases flow must be parallel with the axis of the burner's connector (perpendicular to the burner plane). A uniform distribution of the flow in the transversal section ensures a good operation of the heat recovery steam generator, particularly regarding the superheater. Therefore, the necessary premises are created for ensuring low emissions on the cogeneration group. If the burned gases or the air are uneven distributed, significant variation of the temperatures downstream the burner can occur. Velocity variation in the transversal section, upstream the burner, must not exceed, on 90 % of the burner's section, ± 15 % of the average velocity measured on the entire transversal section. In reality, the burned gases temperature downstream the burner will never be perfectly uniform. Even for a perfect flow distribution of the turbine gases, upstream the burner, the temperature in the area of each burner module will be higher than the temperature between the modules. Therefore, the infrared analysis of the channel connecting the gas turbine and the afterburning installation (silencer – by-pass assembly – connecting transom), at 2xST 18 Plant, has shown unevenness in temperature distribution (figure 17). Considering these phenomena, the afterburning installation can compensate, in good conditions, the mass flow decrease in burned gases produced by the gas turbine at partial loads, keeping a corresponding load on the heat recovery steam generator. In case of turbine stopping, the heat recovery steam generator with the fresh air afterburning is able to keep the steam production at a certain level.

Figure 17. Temperature isotherms, in infrared, in the channel connecting the gas turbine and the afterburning installation (silencer – by-pass assembly – connecting transom)

3.2. Future research

Future research is part of the general context of increasing the flexibility of gas turbine cogeneration groups, the efficiency and reducing the emissions using numerical simulations in CFD environment and experimentations related to: utilization of alternative fuels in gas turbines and afterburning installations, injection of fluids in the cogeneration line in order to reduce the emissions, integrating the gas turbine with fuel cells, etc.

4. Conclusions

Along with the flexibility to alternative fuels feeding, the flexibility of a gas turbine cogeneration plant assumes the accomplishment of several requirements: capability of fast start; capability to pass easily from full load to partial loads and back; maintaining the efficiency at full load and partial loads; maintaining the emission to a low level even when operating on partial loads. Using aeroderivative gas turbines in the cogeneration field has allowed the scientific and technologic knowledge transfer utilization (design concepts, materials, technologies, etc.), which ensures a high degree of energy, from aviation to ground applications. The experience of National Research and Development Institute for Gas Turbines COMOTI Bucharest, in the field of aeroderivative gas turbines (AI 20 GM, TURMO, MK 701, etc.) has allowed the conversion of a gas turbine from liquid fuel to landfill gas, for cogeneration, in stable operating conditions.

Author details

Ene Barbu[1*], Romulus Petcu[1], Valeriu Vilag[1], Valentin Silivestru[1], Tudor Prisecaru[2], Jeni Popescu[1], Cleopatra Cuciumita[1] and Sorin Tomescu[1]

*Address all correspondence to: barbu.ene@comoti.ro

1 National Research and Development Institute for Gas Turbines COMOTI, Bucharest, Romania

2 Politehnica University, Bucharest, Romania

References

[1] Cenusa V., Benelmir R., Feidt M., Badea A. On gas turbines and combined cycles. http://www.ati2001.unina.it/newpdf/Sessioni/Macchine/Impianti/03-Cenusa-Benelmir-Feidt-Badea.pdf (accessed June 5, 2012).

[2] Petcu R. Contributii teoretice si experimentale la utilizarea gazului de depozit ca sursa de energie. Teza de doctorat - Decizie Senat nr. 100/12.02.2010. Universitatea Politehnica Bucuresti; 2010

[3] Carlanescu C. Contributii la problema selectarii si modificarii motoarelor de aviatie pentru utilizarea in scopuri industriale. Teza de doctorat. Universitatea Gheorghe Asachi Iasi; 1994

[4] Energy and Environmental Analysis. Technology Characterization: Gas Turbines. http://www.epa.gov/chp/documents/catalog_chptech_gas_turbines.pdf   (accessed June 6, 2012).

[5] Barbu E., Vilag V., Popescu J., Ionescu S., Ionescu A., Petcu R., Cuciumita C., Cretu M., Vilcu C., Prisecaru T. Afterburning Installation Integration into a Cogeneration Power Plant with Gas Turbine by Numerical and Experimental Analysis. In: Ernesto Benini (ed.), Advances in Gas Turbine Technology. Rijeka: InTech; 2011. p. 139-164. Available from http://www.intechopen.com/articles/show/title/afterburning-installation-integration-into-a-cogeneration-power-plant-with-gas-turbine-by-numerical- (accessed June 6, 2012).

[6] Stationary Sources Branch. Stationary Gas Turbines - 40 CFR Part 60. http://www.cdphe.state.co.us/ap/down/statgas.pdf (accessed June 6, 2012).

[7] Breeze P. Efficiency versus flexibility: Advances in gas turbine technology. PEI 01/04/2011.   http://www.powerengineeringint.com/articles/print/volume-19/issue-3/gas-steam-turbine-directory/efficiency-versus-flexibility-advances-in-gas-turbine-technology.html (accessed May 31, 2012).

[8] Barbu E., Ionescu S., Vilag V., Vilcu C., Popescu J., Ionescu A., Petcu R., Prisecaru T., Pop E., Toma T. Integrated analysis of afterburning in a gas turbine cogenerative power plant on gaseous fuel, WSEAS Transaction on Environment and Development, 2010; 6(6) p. 405-416. http://www.wseas.us/e-library/transactions/environment/2010/89-806.pdf (accessed June 5, 2012).

[9] Jones R., Goldmeer J., Monetti B. Addressing gas turbine fuel flexibility. GE Energy. http://www.ge.com/cn/energy/solutions/s1/GE%20Gas%20Turbine%20Fuel%20Flexibility.pdf (accessed June 6, 2012).

[10] Pimsner V., Vasilescu C., Radulescu G. Energetica turbomotoarelor cu ardere interna. Bucuresti, Editura Academiei RSR, 1964

[11] Marco Antonio Rosa do Nascimento and Eraldo Cruz dos Santos. Biofuel and Gas Turbine Engines, Advances in Gas Turbine Technology. In: Ernesto Benini (ed.), Advances in Gas Turbine Technology. Rijeka: InTech; 2011. p. 116-138. InTech, Available from: http://www.intechopen.com/books/advances-in-gas-turbine-technology/biofuel-and-gas-turbine-engines (accessed June 6, 2012).

[12] Oprea I. Posibilitati de utilizare a gazelor provenite din biomasa in instalatii de turbine cu gaze. ETCN-2005, 30 iunie-1 iulie 2005, Bucuresti, p. 135-139

[13] Jensen J., Jensen A. Biogas and natural gas, fuel mixture for the future. 1st World Conference and Exihibition on Biomass and Energy, 2000, Sevilla. Available from http://www.dgc.eu/pdf/Sevilla2000.pdf (accessed June 11, 2012).

[14] Panoiu P., Marinescu C., Panoiu N., Oroianu I., Mihaescu L. Posibilitati de utilizare a dimetileterului in scopuri energetice. http://caz.mecen.pub.ro/panoiu.pdf (accessed June 11, 2012).

[15] Calin L., Jadaneant M., Romanek A. Gazeificarea biomasei lemnoase. Curierul AGIR, 1-2, ianuarie-iunie 2008, p. 87-90

[16] Chiesa P., Lozza G., Mazzocchi L. Using hydrogen as gas turbine fuel, Journal of Gas Turbine and Power, January 2005, vol. 127 73-80 http://www.netl.doe.gov/technologies/coalpower/turbines/refshelf/igcc-h2-sygas/Using%20H2%20as%20a%20GT%20Fuel.pdf (accessed June 12, 2012).

[17] Ionel I., Ungureanu C., Popescu F. Analiza nivelului de emisii poluante prin schimbarea combustibilui la cuptoarele de tratament termic. http://www.tehnicainstalatii-lor.ro/articole/images/nr12_76-82.pdf (accessed June 14, 2012).

[18] Antonescu N., Polizu R., Muntean V., Popescu M. Valorificarea energetica a deseurilor. Bucuresti. Editura Tehnica; 1988

[19] Ionel P., Borcea Fl., Barbu E., Marinescu C., Ciobanu C. Mihaescu L. Utilizarea combustibililor gazosi regenerabili pentru producerea de energie.Bucuresti. Editura Perfect; 2008

[20] Rainer K. Gas turbine fuel considerations. http://www.scribd.com/doc/76918626/Gas-Turbine-Fuel-Considerations (accessed June 14, 2012).

[21] Fossum M., Beyer R. Co-combustion: Biomass fuel gas and natural gas. http://media.godashboard.com/gti/IEA/ieaCofirNOrep.pdf (accessed June 16, 2012).

[22] Ene M., Ion C., Salcianu R. Cercetari de transformare a unei camere de ardere pentru functionare cu gaze naturale. In: TURBO '98, 13-15 iulie 1998, Bucuresti, Romania

[23] Zubcu V., Zubcu D., Stanciu D., Homulescu V. Instalatie de cogenerare cu componente recuperate, conditii de compatibilitate. in: TURBO '98, 13-15 iulie 1998, Bucuresti, Romania

[24] Komori T., Yamagami N., Hara H. Design for blast furnace gas firing gas turbine. http://www.mnes-usa.com/power/news/sec1/pdf/2004_nov_04b.pdf (accessed June 20, 2012).

[25] Richards G., McMillian M., Gemmen R., Rogers W., Cully S. Issues for low-emission, fuel-flexible power systems. Progress in Energy and Combustion Science 2001; 27: p. 141–169.

[26] Carlanescu C., Manea I., Ion C., Sterie St Turbomotoare – Fenomenologia producerii si controlul noxelor. Bucuresti: Editura Academiei Tehnice Militare; 1998.

[27] Zehe, M.J., Gordon, S. & McBride, B.J. (2002), *CAP: A Computer Code for Generating Tabular Thermodynamic Functions from NASA Lewis Coefficients*, NASA Glenn Research Center, NASA TP—2001-210959-REV1, Cleveland, Ohio, U.S.A., http://www.grc.nasa.gov/WWW/CEAWeb/TP-2001-210959-REV1.pdf (accessed June 26, 2012).

[28] Corna N., Bertulessi G. The use of oxigen in biomass and waste-to-energy plants: A flexible and effective tool for emission and process control, Third International Symposium on Energy from Biomass and Waste, 8-11 November 2010, Venice, Italy

[29] Drnevich R., Meagher J., Papavassiliou V., Raybold T., Stuttaford P., Switzer L., Rosen L. Low NO_x emissions in a fuel flexible gas turbine, Issued August 2004, http://www.netl.doe.gov/technologies/coalpower/turbines/refshelf/reports/41892%20Praxair%20Final%20Report_Low%20NOx%20Fuel%20Flexible%20Gas%20Turbine.pdf (accessed June 26, 2012).

[30] Daiber J., Fluid dynamics of the HRSG gas side, Power, March 2005, p. 58-63 http://www.babcockpower.com/pdf/vpi-45.pdf (accessed June 26, 2012).

Chapter 5

Development of Semiclosed Cycle Gas Turbine for Oxy-Fuel IGCC Power Generation with CO$_2$ Capture

Takeharu Hasegawa

Additional information is available at the end of the chapter

1. Introduction

In response to recent changes in energy-intensive and global environmental conditions, it is urgent and crucial concern to develop the high-efficiency technologies of fossil fuel power generations. Especially, coal is one of the most important resources from the standpoint of risk avoidance in the scheme of power supply composition. Figure 1 shows the proved recoverable reserves of coal by region compared with those of the natural gas and crude oil. The world's coal reserves are twice that of each conventional oil and natural gas, distributed more evenly on a geographical basis than those for oil and natural gas, and also geopolitical risk is lower for securing the stable supply of coal resource. This figure also shows each total discoverable reserve of non-conventional resources of natural gas and crude oil as references, and each reserve corresponds to twice of the coal proved recoverable reserve. In this regard, however, total discoverable reserve of coal is estimated ten times of proved recoverable reserves, or it is corresponds to five times of that of each non-conventional resource of natural gas and crude oil. Coal is definitely the most important fossil fuel resources in the future.

Furthermore, in the 1997 when the Third Conference of Parties to the United Nations Framework Convention on Climate Change (COP3), the Kyoto protocol, which invoked mandatory CO$_2$ emissions reductions on countries, was adopted. CO$_2$ emissions per unit calorie of coal are about 1.8 times that in the case of natural gas, and then CO$_2$ recovery technologies are very important for thermal power plants.

On the other hand, demand of coal has increased rapidly in the recent years. Figure 2 shows annual changes of the world's coal consumption by region and the reserves-to-production ratios of coal, oil and natural gas. In the intervening quarter-century from 1985 to 2010, the coal consumption in Asia Pacific increased significantly or about 3.6 times, while world coal consumption increased 1.7 times. The increase in coal consumption in Asia Pacific is equal to one half of the world's consumption in 2010, while consumptions in other regions decrease. In just ten years, coal consumption in Asia Pacific increased double, and then the world's reserves-to-production ratio of coal decreased by half, while the reserves-to-production ratios of oil and natural gas have been maintained constant. Along with the growing world demands for fossil energy resources in recent years, international competition for development of fossil fuel fields of coal, oil and gas in the world is ever intensified.

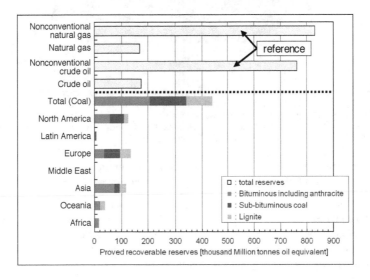

Figure 1. Proved recoverable reserves of coal by region at end 2010, compared with oil and natural gas reserves. Source of reserves data: BP statistical review of world energy 2011 [1]. Notes: Coal proved reserves expressed in tonnes oil equivalent are calculated using coal productions based on data expressed in tonnes oil equivalent and coal productions in tonnes. Nonconventional natural gas shows data not including methane hydrate reserves. Nonconventional crude oil includes oil shale and oil sand reserves.

With the above mentioned situations as a background, developments of high-efficiency power generation technologies and low emission technologies of CO_2 become increasingly important in the world. As one of the highly-efficient and low CO_2 emission technologies, an integrated coal gasification combined cycle (IGCC) power generation combined with CO_2 capture and storage (CCS) technologies are now drawing attention from the electric power industry. The Central Research Institute of Electric Power Industry (CRIEPI) has proposed a newly-designed oxy-fuel IGCC power generation system integrated with a combination of CO_2 recovery processing and a semiclosed cycle gas turbine [3]. This system wields the advantages of not requiring a CO_2 capture system using CO_2 absorption processing or fuel reforming preprocessing. Compared to conventional CO_2 recovery thermal power plants, oxy-fuel IGCC could simplify CO_2 recovery systems, reduce station service power, and achieve higher thermal efficiency. Currently, CRIEPI is addressing each technological development

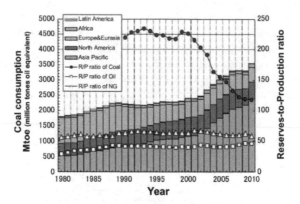

Figure 2. World coal consumption by region and proved recoverable reserves-to-production (R/P) ratio of coal, oil and natural gas (NG) at end 2010. Note: Coal data include anthracite, bituminous, sub-bituminous, and lignite. And reserves-to-production (R/P) ratios are approximate values based on the total proved recoverable reserves of bituminous coal, anthracite, lignite and sub-bituminous coal. Sources are BP statistical review of world energy [1] and data reported for precious World Energy Council Surveys of Energy Resources [2].

[4-9] towards the realization of highly efficient power generation with zero emissions, and with a semiclosed gas turbine system serving as one of the key technologies.

In this study, we have been researching and developing the combustion technologies in order to achieve the semiclosed cycle gas turbine for highly efficient oxy-fuel IGCC [5-6]. This paper describes technical difficulties and combustion characteristics of semiclosed gas turbine combustors, comparing developed H_2/O_2 and natural gas/O_2 fired semiclosed gas turbines in the WE-NET project [10] and a conventional natural gas fired gas turbine.

2. CO_2 recovery from thermal power plant

2.1. CO_2 recovery methods for IGCCs

Along with the oxy-fuel IGCC system newly proposed in this paper, there exist four CO_2 recovery systems for coal-base thermal power generation. With regard to CO_2 recovery systems for IGCC, as shown in figure 3, the oxy-fuel IGCC system and the pre-combustion system for IGCC are under development [11-14]. In the case of an oxy-fuel IGCC power generation system with CO_2 capture in a semiclosed cycle oxy-fuel gas turbine, recovery of CO_2 is simplified, with decreasing station service power expected to produce highly efficient generation. This is because water-gas-shift reactors and physical/chemical solvents for CO_2 capture are not required as opposed to conventional pre-combustion systems for IGCC.

Figure 4 shows the change in net plant efficiency (HHV basis) in conventional pulverized coal and IGCC power plants. In the case of 90 percent CO_2-recovery, post-combustion systems, the thermal efficiency of pulverized-coal, super critical boilers decreases to 28.4% [11]

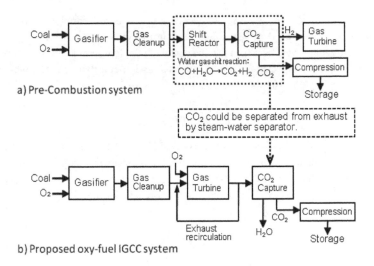

a) Pre-Combustion system

b) Proposed oxy-fuel IGCC system

Figure 3. Comparison of CO₂ recovery processes for IGCCs

from 39.3% since a huge amount of steam is needed to regenerate absorbers, while oxy-fuel combustion systems of O_2-fired pulverized coal boilers result in only a marginal improvement in thermal efficiency of 29.3% [11]. Furthermore, in the case of a pre-combustion system using an F-class gas turbine for IGCC, thermal efficiency is expected to improve to 31.6% [11].

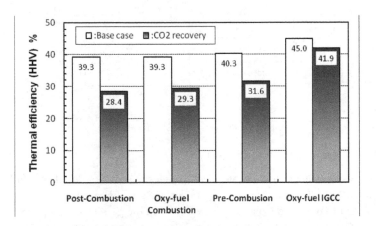

Figure 4. Thermal efficiency of coal-base power plants with and without CO₂ capture and compression. In the three conventional cases of post-, oxy-fuel and pre-combustion, currently available technologies are employed and CO₂ recovery rate is set at 90% [11]. In the case of oxy-fuel IGCC employing technologies currently under development, CO₂ recovery rate is set at 99% [4].

However, in the case of an oxy-fuel IGCC system adopting each technology currently under development, the use of an O_2-CO_2 gasifier, for example, with a hot/dry synthetic gas clean-up system and a semiclosed cycle gas turbine (turbine inlet temperature on ISO standard basis at about 1530K), is expected to produce a transmission-end thermal efficiency of 41.9% under conditions of 99% or higher CO_2 recovery.

2.2. Oxy-fuel IGCC and closed-cycle gas turbine

Figure 5 shows a schematic diagram of the oxy-fuel IGCC system and a topping semiclosed cycle gas turbine. The newly proposed oxy-fuel IGCC consists of an oxygen-CO_2 blown gasifier, a hot/dry synthetic gas cleanup system, a semiclosed cycle oxygen-fired gas turbine, and a CO_2 recovery process. This system has the following advantages;

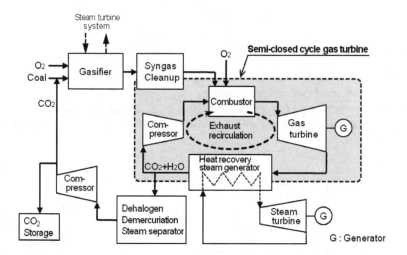

Figure 5. Schematic diagram of oxy-fuel IGCC and semiclosed cycle gas turbine [3]

- Oxygen-CO_2 blown, entrained-flow coal gasifier

Table 1 shows the rated conditions of a gasified fuel and semiclosed cycle gas turbine combustor [3],[4]. Table 2 shows characteristics of coal used in the calculation [3]. Here, we dry fed pulverized coal into an oxygen-blown entrained-flow gasifier with recycled CO_2 from flue gas, and gasified with additional oxygen. In addition, we found that O_2-CO_2 blown coal gasification enhanced gasification efficiency compared to that of current oxygen blown gasification through dry feeding of coal with N_2. Figure 6 shows the gasification characteristics of the two cases above, estimated by numerical analysis of a one-dimensional model [3].

Components	Gasified fuel	Oxidizer	Dilution
CO [vol%]	66.2	0	0
H_2	23.8	0	0
CH_4	0.3	0	0
CO_2	4.9	0	69.5
H_2O	3.2	0	26.9
Ar, N_2	1.5	2.5	2.7
O_2	0	97.5	0.9
HHV (LHV)	11.5 MJ/m³ (11.0 MJ/m³) at 273K, 0.1MPa		
Pressure in combustor	2.2MPa		
ϕ *	0.98 (Overall equivalence ratio ϕ is 0.89)		
Dilution ratio	5.5: dilution/fuel molar ratio		
Exhaust temp.	1573K at combustor exit		
ϕ* : calculated from fuel and oxidizer without O_2 concentration in dilution			

Table 1. Rated conditions of semiclosed cycle gas turbine combustor [3],[4]

Inherent moisture* [wt%]	3.6
Ash content* [wt%]	9.6
Volatile matter* [wt%]	30.3
Fixed carbon* [wt%]	56.5
Ultimate analysis**	
C [wt%]	76.1
H [wt%]	5.1
O [wt%]	6.9
N [wt%]	1.7
S [wt%]	0.5
*: air-dried state, **: dry basis	

Table 2. Characteristics of coal used in calculation [3]

Table 3 shows numerical analysis conditions in gasification. Gasified fuels were calculated under conditions where an equivalence ratio in the gasification was set at 2.58 through multi-stage analyses utilizing pyrolysis, char gasification reaction and gas phase equilibrium reaction processes, and assuming a one-dimensional axial flow in the entrained-flow gasifier [3].We assumed that volatile matter contents in coal would be instantaneously pyrolyzed in

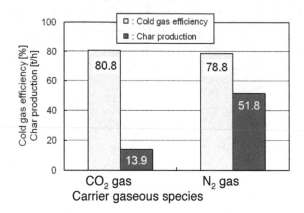

Figure 6. Influence of carrier gas conveying pulverized coal into gasifier on oxygen-blown gasification performance under conditions of coal input of 118.5t/h [3]

the first stage, so we took 3-step reduced reactions in char gasification into account. For char gasification, we used char reaction rates based on experimental data from a pressurized drop tube furnace [15]. In the analyses, we determined the point in time when char input accorded with char production to be equilibrium. Since we assumed 100% removal rates of dust and sulfur in the synthetic gas cleanup, gasified fuels shown in Table 1 did not include sulfur, halide, ash and metal impurities.

Calculation	One-dimensional model
Reaction	
1) Pyrolysis	
Coal → $C_nH_mO_l$ + Char	Pyrolyzed instantaneously
2) Reaction of Char	
$C + 1/2O_2 → CO$	Reaction rates obtained from data in
$C + CO_2 → 2CO$	pressure drop tube furnace
$C + H_2O → CO+H_2$	
3) Gas phase reaction $(C,H,O) → (CH_4, H_2, CO, CO_2, H_2O, N_2, O_2)$	Equilibrium reaction

Table 3. Analysis method and conditions [3]

The cold gas efficiency in Fig.6 demonstrates the ratio between chemical energy content in the product gas compared to chemical energy in fuel on a lower heating value basis. Cold gas efficiency was calculated in the following way:

$$cold\,gas\,efficiency = \frac{product\,gas\,[mass\,flow\times heating\,value]-additional\,fuel\,[mass\,flow\times heating\,value]}{coal\,sup\,plied\,to\,gasifier\,[mass\,flow\times heating\,value]}\times 100 \qquad (1)$$

As a result, we estimated an improvement in cold gas efficiency by 2 percent and a reduction of char particles. At the same time, we clarified the influence of CO_2 and H_2O content on char production characteristics by using a pressurized drop tube furnace [8], and we evaluated the effects of CO_2 enrichment on coal gasification performance using an actual pressurized entrained flow coal gasifier of a 3ton/day bench scale gasifier [9]. Results confirmed that CO_2 enrichment improves gasification characteristics.

• Hot/dry synthetic gas cleanup

We treated gasified fuels with a hot/dry synthetic gas cleanup system consisting of a metallic filter, a hot gas desulfurization unit and other materials, which simplified the cleanup system and reduced the power consumption for cleanup [7]. Dust removal technologies using metallic filters or ceramic ones have already been demonstrated and put to practical use in IGCC plants. So far, the Central Research Institute of Electric Power Industry has developed a halide sorbent containing $NaAlO_2$ [16], a honeycomb zinc ferrite desulfurization sorbent containing ZnFe2O4 [7], a honeycomb copper based mercury sorbent containing CuS [17], and an ammonia decomposing Ni-based catalyst supported by ZSM-5 pellets [18] and each of those elemental technologies was expected to be applied to the hot/dry synthetic gas cleanup system for current IGCCs. Figure 7 [19] shows the schematics of the demonstration plant of the dry gas purification system for the IGCC now being developed. An ammonia catalytic removal process was expected to be installed following the desulfurization unit. The process sequence of the purification system was determined by considering the operation temperature and performance of the sorbents and catalyst. Recently, the Central Research Institute of Electric Power Industry has been moving ahead on design of a new dry gas purification system for the advanced oxy-fuel IGCC by applying the purification system employing the elemental technologies developed for current IGCCs. Impurities in gasified fuels such as dust, ash contents, metal compounds, sulfur, halide, mercury and others could be reduced to an allowable level [20] for conventional gas turbines.

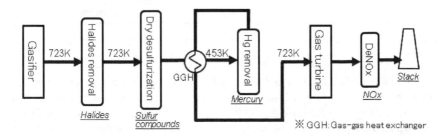

Figure 7. Schematic flow diagram of demonstration plant of dry gas purification system for current IGCCs [19]

• Semiclosed cycle gas turbine and CO_2 recovery

In a semiclosed cycle oxy-fuel gas turbine system as a topping cycle, we burned gasified fuels with pure oxygen and adjusted combustor exhaust temperature by recycling CO_2-enriched flue gas. As shown in Table 1, the rated temperature of combustor exhaust was set at 1573K (1300degC) and pressure inside the combustor at 2.2MPa [4]. After recovering exhaust heat in the HRSG, the necessary amount of flue gas was compressed and recycled to a gas turbine.

We then fed the remaining flue gas to a water scrubber of a halogen and Hg removal system and mist separator. We found that following these treatments, flue gas consisting mostly of CO_2 and H_2O became high-concentration CO_2 gas. We used some of the flue gas to feed coal to a gasifier, with the remainder compressed and sent to a storage site. It was necessary to reduce oxygen concentration in coal carrier gas to a low level in order to prevent pulverized coal from firing inappropriately.

Table 4 shows subjects and characteristics of gasified fuel/O_2 stoichiometric combustion with exhaust recirculation compared to a conventional natural gas-fired gas turbine. Unlike in the case of excess air combustion of an natural gas-fired gas turbine, the suppression of fuel oxidation under O_2-fired stoichiometric conditions with exhaust recirculation poses concerns, thereby necessitating the development of combustion promotion technology.

	Oxy-fuel combustion in IGCC	Conventional natural gas-fire GT
Equivalence ratio	Stoichiometric (0.98)	0.4~0.5
	Oxidation reaction is restrained and unburned fuel is emitted.	at T_{ex}=1573K ~1773K
Dilution gas to adjust combustion temp.	Exhaust recirculation	Air
	Some exhaust is used as coal carrier gas, and then O_2 concentration has to be decreased to a safe level.	
NOx emissions	Hot/dry cleanup and exhaust recirculation cause increased NOx emissions	Only thermal-NOx emissions

Table 4. Subjects of semiclosed cycle gas turbine of gasified fuel/O_2 stoichiometric combustion with exhaust recirculation

In the case of oxy-fuel combustion in IGCC, a little excess O_2 combustion in which apparent equivalence ratio is set at 0.98 or lower resulted in higher concentrations of residual O_2 in exhaust, restraining the usage of exhaust to feed coal into the gasifier while combustion efficiency rose. And the presence of non-condensable gases such as remaining O_2, and Ar and N_2 separated from the air resulted in increased condensation duty for the recovery of the CO_2 [21]. On the other hand, a little higher equivalence ratio over stoichiometric conditions decreased combustion efficiency. We have to accomplish higher combustion efficiency under almost stoichiometric conditions and decrease.

Furthermore, both the employment of hot/dry synthetic gas cleanup and exhaust recirculation increased fuel-NOx emissions.

Against the above backdrop, we first of all researched combustion characteristics and exhaust gas reaction characteristics in the semiclosed cycle gas turbine for oxy-fuel IGCC [5].

3. Numerical analysis method based on elementary reaction models with PSR and PFR

We examined the reaction characteristics of reactant gases both in the combustor and in exhaust using numerical analysis based on the following elementary reaction kinetics. Here, we employed the reaction model proposed by Miller and Bowman [22], and confirmed by test result comparison the appropriateness of the model for non-catalytic reduction of ammonia in gasified fuel using NO [23] and an oxidation of ammonia by premixed methane flame [22].

The reaction scheme we employed was composed of 248 elementary reactions, with 50 species taken into consideration. Miller and Bowman described both a detailed scheme of the oxidation of C1 and C2 hydrocarbons under most (but not too fuel-rich) conditions, and an essential scheme for ammonia oxidation. Hasegawa et al. [23], united these two schemes and confirmed the applicable scope of a united scheme through experiments using a flow tube reactor. As an example, figure 8 shows comparative calculations results with non-catalytic denitration tests performed by Lion [24]. The analytical results precisely described a narrow reaction temperature for effective non-catalytic denitration and the behavior of NH_3 and NO constituents. Furthermore, the authors have evaluated the reaction characteristics of ammonia reduction in the gasified fuels, of non-catalytic denitration in exhaust, of air-fired gasified fueled combustions, and of H_2/O_2 stoichiometric combustion through experiments and full kinetic analyses [23], [25]-[27]. Results showed that the united scheme could describe the reaction characteristics in gasified fueled combustion and exhaust. On the other hand, various reaction schemes have been proposed worldwide for each reaction system including higher hydrocarbons. There was example of the GRI Mech 3.0 chemical kinetic mechanism used for calculation of the oxy-fuel gas turbine combustion [28]. But it need not be used since the gasified fuel contains a small percent of CH_4 and no C2 hydrocarbon.

We took thermodynamic data from the JANAF thermodynamics tables [29], and calculated the values of other species not listed in the tables based on the relationship between the Gibbs' standard energy of formation, $\Delta G°$, and the chemical equilibrium constant, K, obtaining a value of $\Delta G°$ from the CHEMKIN database [30].

$$\Delta G° = R \times T \times \ln(K) \tag{2}$$

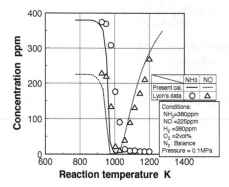

Figure 8. Comparison of kinetic analyses with experimental data of Lyon [24] on concentration of NH₃ and NO in the NH₃NOO₂H₂ system under conditions of selective noncatalytic reduction of NOx

We in this study used the GEAR method [31] for numerical analysis as an implicit, multistage solution.

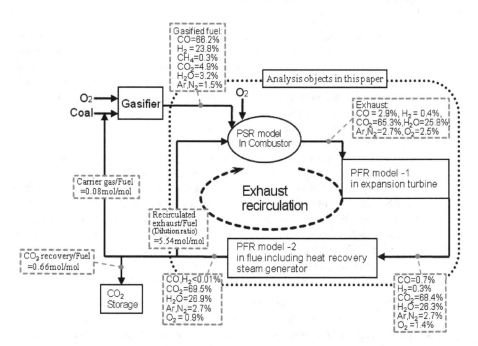

Figure 9. Schematic of algorism in semiclosed gas turbine for oxy-fuel IGCC under typical rated conditions. Recirculated gas turbine exhaust is injected into the PSR of the combustor alongside incoming stream of gasified fuel and oxidizer of O_2.

Furthermore, our algorithm is schematized in Figure 9. The model we employed in this study assumed all mixing processes to be ideal such that they could be represented by a combination of a perfectly-stirred reactor (PSR) and a plug flow reactor (PFR). When investigating the basic combustion reaction characteristics that were independent of combustor geometries, the combustor was simply modeled as the PSR. This combustor model was the simplest case of modular models employed by Pratt, et al. [32]. In the case of investigating the exhaust gas reaction characteristics in expansion turbine and flue, we employed the PFR model. Then, we employed a combination PSR and PFR model in order to explore the influence of exhaust recirculation on combustor emission characteristics and exhaust reaction characteristics in the semiclosed gas turbine.

4. Characteristics of stoichiometric combustion with recirculating exhaust

4.1. Comparison with air-fired combustion

Figure 10 shows concentrations of principal chemical species against reaction time under the rated load conditions shown in Table 1, through a numerical analysis based on reaction kinetics with a PSR model of homogeneous reaction. Figure 11 also shows the principal chemical species against reaction time when burning CH_4 in the main components of natural gas with air under conditions where the reaction temperature is set at a constant value of the rated exhaust temperature of 1573K, and where the equivalence ratio is 0.32.

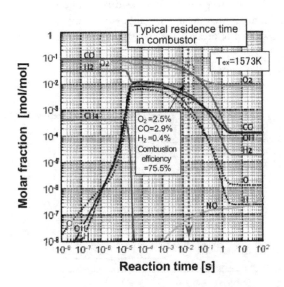

Figure 10. Chemical species behavior over time in gasified fuel/O_2 stoichiometric combustion with exhaust recirculation

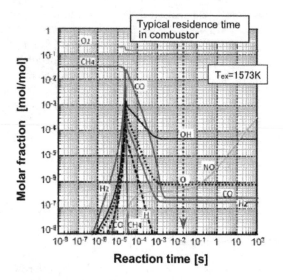

Figure 11. Chemical species behavior over time in conventional CH₄/air combustion

In the case of burning gasified fuel under stoichiometric conditions with exhaust recirculation, fuel oxidation reaction proceeded slowly compared to that of conventional CH₄/air
combustion. As a result, we found that CO and H₂ in exhaust remained unoxidized at
around 2.9vol% and 0.4vol%, respectively, and residual O₂ at 2.5vol% in 20ms corresponded
to the combustion gas residence time in the combustor. Combustion efficiency was estimated to remain at a low level of around 76%, compared with that of conventional industrial
gas turbines.

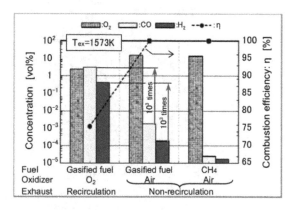

Figure 12. Comparison of emission characteristics with conventional air-fired combustions

Figure 12 shows exhaust characteristics and combustion efficiencies at the combustor exit under the conditions of a 1573K combustor exhaust temperature, in comparison with the homogeneous premixed combustion of a gasified fuel/air and CH_4/air mixture. The stoichiometric combustion of gasified fuel/O_2 with exhaust recirculation causes a drastic decrease in combustion efficiency compared with the other two cases of air-fired combustion. We feel that it is therefore necessary to promote fuel oxidation, or to decrease combustible constituent CO emitted from the gas turbine.

4.2. Comparison to each oxy-fuel gas turbine combustion

Figure 13 shows the results of numerical analysis in hydrogen/oxygen fired, stoichiometric combustion with exhaust recirculation of steam under the rated temperature conditions of 1573K. An overall equivalence ratio was set at 1 with other conditions equivalent to the rated conditions in Table 1.

Hydrogen/oxygen reaction began as rapidly as in the cases of gasified fuel/O_2 or CH_4/air fired combustion, shown in Fig.10 or Fig.11, respectively. After that, hydrogen oxidation reaction progressed faster than in the case of gasified fuel/O_2 fired combustion.

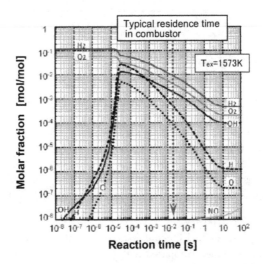

Figure 13. Chemical species behavior over time in H_2/O_2 combustion with steam recirculation

Figure 14 shows exhaust characteristics and combustion efficiencies under the conditions of a 1573K combustor exhaust temperature compared with the cases of homogeneous premixed combustions of H_2/O_2 and CH_4/O_2 mixture with exhaust recirculation. Here, we set the composition of each recirculating exhaust to that of corresponding gas formed under equilibrium conditions.

As an example of oxygen-fired gas turbine using stoichiometric combustion with exhaust re-circulation, Fig.14 also shows comparative calculations with test data of 1973K-class H_2/O_2 stoichiometric combustion with steam recirculation, conducted in the Japanese WE-NET project. Tests confirmed that the analytical results were almost in accordance with experimental results [33] concerning concentrations of residual O_2 constituent and unburned H_2 constituent in exhaust, and that the numerical analysis used in this study could estimate emission characteristics under conditions of achieved high combustion efficiency.

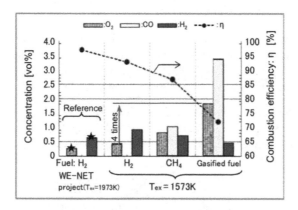

Figure 14. toichiometric combustion characteristics of each fuel; overall equivalence ratio is 1, markers (⋆)are test data under conditions where combustion pressure is set at 2.5MPa and recirculated steam temperature is 623K [33].

In the case of H_2/O_2 fired combustion such as in a hydrogen fired closed-cycle gas turbine, emissions of combustible constituent H_2 and residual O_2 in exhaust decreased 1vol% or below in a reaction time of 20ms, or combustion efficiency was estimated to reach up to around 93% at a temperature of 1573K. In the case of a CH_4/O_2 fired, closed cycle gas turbine, combustible constituent CO and residual O_2 concentration of combustor exhaust also decreased 1vol% or below, and combustion efficiency reached 87%, while combustible contents and residual O_2 emissions displayed a tendency to increase compared to the H_2/O_2 fired combustion.

Combustible contents and residual O_2 emissions, on the other hand, increased by several times in the case of gasified fuels compared with both cases of H_2 fired and CH_4 fired combustion. Combustion efficiency fell to a low level of 72%. In CO-rich fuel/O_2-fired combustion with exhaust recirculation, CO oxidation was strongly restrained by recirculating exhaust consisting mostly of CO_2 compared to other fuel constituents, and combustion efficiency was decreased. Therefore, to achieve highly efficient oxy-fuel IGCC, it is necessary to develop combustion control technologies of gasified fuel/O_2 combustion with higher combustibility compared with the H_2/O_2 combustion technology in the WE-NET project or pre-combustion technologies.

4.3. Effects of fuel CO/H_2 molar ratio on emission characteristics

Each quantity of CO and H_2 constituent in the gasified fuels differs chiefly according to the gasification methods, raw materials of feedstock, and water-gas-shift reaction as an optional extra for pre-combustion carbon capture system. Figure 15 shows influences of CO/H_2 molar ratio in the gasified fuel on the combustion emission characteristics with exhaust recirculation under the rated temperature condition of 1573K. In the case of varying the fuel CO/H_2 molar ratio under the conditions where the total amount of CO and H_2 constituent was set constant, dilution ratio (dilution gas/fuel molar ratio) was adjusted to maintain the combustion temperature at 1573K. Just like the case of Fig.14, overall equivalence ratio was set at 1, with other conditions equivalent to the rated conditions in Table 1. In the case of changing the fuel CO/H_2 molar ratio from 2.8 of base condition to 0.36, the amounts of CO and H_2 constituent replaced each other under the condition where the total amount of CO and H_2 was set constant of 90vol%.

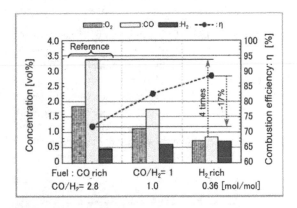

Figure 15. Effects of CO/H_2 molar ratio in fuel on stoichiometric combustion characteristics; overall equivalence ratio is 1. Notes: In the case of changing the fuel CO/H_2 molar ratio from 2.8 of base condition to 0.36, the amounts of CO and H_2 constituent replace each other under the condition where the total amount of CO and H_2 is set constant of 90vol%.

In the case of higher CO/H_2 molar ratio in the fuel, higher concentration of CO and lower concentration of H_2 in fuels increased CO emissions in combustion exhaust significantly, but have insignificant effects on reduction of H_2 emissions. As a result, in the case of CO rich gasified fuels, CO emissions increased four times those in the case of H_2 rich gasified fuel in the pre-combustion IGCC system, or combustion efficiency decrease by about 17%. This is explained both because H_2 is decomposed and produces OH, H and O radicals in the chain initiation as shown in Fig.10, and exhaust recirculation strongly inhibits oxidation of CO that is oxidized directly to CO_2 by the following reactions:

$$CO + OH\left(O + M, O_2, HO_2\right) \Leftrightarrow CO_2 + H\left(M, O, OH\right), \quad M\text{:Thirdbody} \tag{3}$$

Furthermore, H_2 is oxidized more rapidly than CO, or CO constituent controls overall oxidation reaction rate of fuel in the stoichiometric combustion with exhaust recirculation. Consequently, when the CO/H_2 molar ratio increased, CO oxidation rate and O_2 consumption rate decreased.

4.4. Effects of equivalence ratio on emission characteristics

Figure 16 shows the effects of an equivalence ratio on combustion emission characteristics under the rated temperature 1573K. When varying the equivalence ratio, the dilution ratio (dilution gas/fuel molar ratio) was adjusted to maintain the combustion temperature at 1573K. The horizontal axis indicated an apparent equivalence ratio, φ^* calculated from fuel and an oxidizer without O_2 concentration in the dilution of recirculated exhaust. Emission features and combustibility of the combustor were characterized by combustion conditions near the burner. That is, Fig.16 indicated the influence of a so-called "local equivalence ratio" near the burner on combustion emission characteristics by using the apparent equivalence ratio φ^*.

In the case of decreasing φ^* from 0.98 to 0.95, combustion efficiency increased by only 5 percent, while overall equivalence ratio decreased from 0.89 to a low level of 0.75. That is, lowering the equivalence ratio could not result in any remarkable combustion promotion in CO-rich fuel/O_2 fired combustion with exhaust recirculation, while O_2 concentration in the exhaust significantly increased and the usage of exhaust to feed coal into the gasifier was restrained. It is necessary to decrease O_2 concentration in the carrier gas to feed coal by oxidation reactions using fuels such as hydrocarbons, or auxiliary power increased. Therefore, we have to decide the equivalence ratio in the combustor in consideration of the influence of residual O_2 on thermal efficiency of the whole system and performance of its equipments.

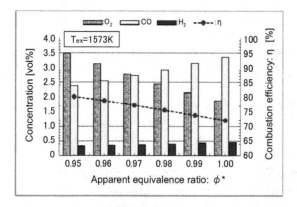

Figure 16. Effects of apparent equivalence ratio(φ^*) on combustion emission characteristics

4.5. Influences of oxygen concentration in oxidizer on emission characteristics

O_2 concentration in oxidizer derived from an air separation unit differs according to the air separation and purification system. Figure 17 shows influences of oxygen concentration in oxidizer on the combustion emission characteristics under the rated temperature conditions. In the case of varying the oxygen concentration in oxidizer, dilution ratio (dilution gas/fuel molar ratio) was adjusted to maintain the combustion temperature at 1573K and overall equivalence ratio at 1.0. The remainder of the oxidizer without O_2 was set to N_2.

Emissions of residual O_2 and combustible constituents of CO and H_2 in exhaust tended to increase with the increase in oxygen concentration in oxidizer, or combustion efficiencies decreased. In the case of increasing O_2 concentration from 80vol% to 100vol%, combustion efficiency decreased by 13%, while residual concentrations of argon and nitrogen originated in air decreased. It was said that the non-condensable gases such as remaining O_2, argon and nitrogen resulted in increased condensation duty for the recovery of the CO_2 [21], or influence of residual constituents on the whole system and its equipments must be examined separately.

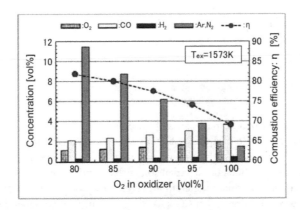

Figure 17. Effects of oxygen concentration in oxidizer derived from air separation unit on combustion emission characteristics; overall equivalence ratio is 1.

4.6. Reaction characteristics of gas turbine exhaust

Figure 18 shows a typical stream history of exhaust temperature and pressure from a gas turbine inlet to a compressor inlet of recirculating exhaust. Power was recovered from exhaust emanating from the combustor in the expansion turbine, and combustor exhaust temperature of 1573K with a pressure of 2.2MPa decreased to around 950K and 0.1MPa respectively at the turbine exit. Then, heat from expansion turbine exhaust was recovered through heat recovery steam generator (HRSG) in a flue, and exhaust temperature decreased to around 373K at the compressor inlet. In these analyses, we employed the PFR model for the turbine exhaust to the compressor inlet, and assuming that species in exhaust is evenly mixed and that there is no distribution of temperature and pressure in the mixtures. There is also no supply of added oxidizer and recirculating exhaust in the reaction processes.

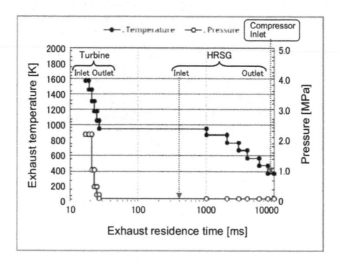

Figure 18. Typical stream history of exhaust temperature and pressure from gas turbine inlet to compressor inlet of recirculating exhaust

Figure 19 shows the reaction characteristics of combustion gas in the combustor and exhaust gas from combustor outlet to compressor inlet of recycled exhaust using a combination PSR and PFR model. CO and H_2 at high concentration in exhaust could be slowed to oxidize under the temperature conditions of an expansion turbine and HRSG. Combustible constituents of CO and H_2, and residual O_2 therefore decreased in concentration along the exhaust flow direction. Oxidation reactions of CO and H_2 then nearly halted when the exhaust temperature decreased to 673K or less.

Figure 20 shows emission characteristics of exhaust gases and combustion efficiencies at typical conditions based on the above reaction characteristics.

In a reaction time of 400ms corresponding to the residence time between combustor exit and HRSG inlet, combustible constituent CO and H_2 decreased less than 0.01vol%, and residual O_2 decreased to around 0.9vol%, while CO and H_2 concentration in combustor exhaust hovered at around 3vol% and 0.4vol%, respectively, and residual O_2 was at 2.3vol% under 20ms of typical combustion gas residence time in the combustor. Each oxidation reaction of combustible constituents in the turbine and the flue resulted in an increase in combustion efficiency of η by about 12%, respectively, or a combustion efficiency was estimated to reach a high level of around 99.8%. If the reaction time was from 4 to 10 seconds when the exhaust

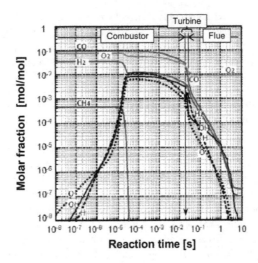

Figure 19. Chemical species behavior of combustion and exhaust gas over time in semiclosed cycle gas turbine, using PSR + PFR combined model. Combustor inlet conditions are the same as those in Fig.10 and flue includes HRSG.

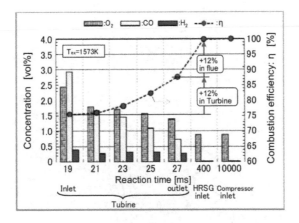

Figure 20. History of combustion emissions from gas turbine inlet to compressor inlet of recirculating exhaust

gas temperature decreased to around 673K in the HRSG, both combustible constituent CO and H_2 decreased to 10ppmv, or the combustion efficiency reached 100%.

From the abovementioned results, we were able to clearly slow combustible constituents in expansion turbine exhaust to oxidize under conditions of exhaust temperatures of over 673K, or recover burning energy from unburned fuel in HRSG. However, in this case, since reaction heat of combustible constituents in HRSG corresponds to a fuel for reheat type HRSG, some of the supplied fuel could not devote enough energy to a combined cycle thermal efficiency. In order to achieve highly efficient oxy-fuel IGCC, it was therefore necessary to increase combustion efficiency as much as possible in the gas turbine combustor.

4.7. Influences of exhaust recirculation on thermal-NOx emissions

As shown in Fig.20, it is found that combustible constituents reached almost equilibrium concentration at compressor inlet of recirculated exhaust, or that equilibrium gases were working fluids in the semiclosed cycle gas turbine. On the other hand, NOx constituents increased by exhaust recirculation and were saturated by a balance between exhaust recirculation and CO_2 recovery process. Figure 21 demonstrates the influence of exhaust recirculation on thermal-NO emission characteristics through numerical analyses of a combination PSR and PFR model as in the case of Fig.20, with compositions of fuel and oxidizer shown in Table 1. In these analyses, Fig.21 indicates the direct effects of recirculating NO constituent on NO-saturating concentration under conditions where dilution gas composition without NO constituent is constant. That is, we repeated the calculation of one exhaust-recirculating loop shown in Fig.20, and investigated influence of exhaust recirculation on oxidation-reduction reaction of NO through full kinetic analyses. Combustion temperature was set at 1623K, and pressure at 3.0MPa; a little higher than indicated in Table 1. Dilution gas of recirculated exhaust at combustor inlet were set to equilibrium composition.

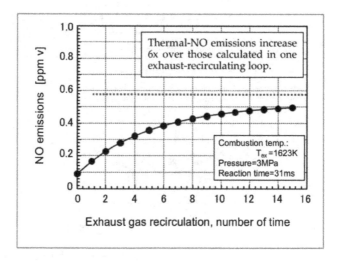

Figure 21. Influence of exhaust recirculation on thermal-NO emissions

Thermal-NO emissions increased in response to a number of times exhaust was recirculated and reached around 6 times higher than those calculated in one exhaust-recirculating loop, while NO production itself was not significantly large due to the small component amounts of N_2, as shown in Table 1. However, since thermal-NO production depends on reaction temperature, or thermal-NO emissions are strongly affected by both mixing processes and non-uniformities of mixtures, we need further studies on thermal-NO emissions in the processes of combustion and combustor design.

5. Conclusions

Oxy-fuel IGCC employing an oxygen-fired, semiclosed cycle gas turbine with exhaust recirculation enables the realization of highly-efficient, zero-emissions power generation. Numerical analyses in this paper showed both combustion emission characteristics of the semiclosed cycle gas turbine combustor and oxidations of unburned fuel constituents in the turbine exhaust in a flue, compared with conventional air-fired gas turbines and advanced O_2-fired gas turbines. As a result, we were able in this study to clarify that unburned constituents in combustor exhaust were slow to oxidize under temperatures of over 673K in the flue and that all fuel energy could be used for power generation, while the oxidation reaction of CO-rich gasified fuel under stoichiometric conditions could be restrained with CO_2 constituents in re-circulated exhaust at decreased combustion efficiency. In this case, however, all the supplied fuel could not devote enough energy to boosting combined cycle thermal efficiency, leading therefore to a decrease in thermal efficiency overall. As a next step, we propose the need to promote oxidation reaction by developing combustion control technology for the improvement of plant thermal efficiency.

Nomenclature

Dilution ratio: dilution gas of exhaust recirculation over fuel supply ratio, [mol/mol]

HRSG: heat recovery steam generator

T_{ex}: average temperature of combustor exit gas, [K]

η: combustion efficiency, [%]

$$combustion\,efficiency = (1 - \frac{combustor\,exhaust\,gas[mass\,flow \times lower\,heating\,value]}{gasified\,fuel\,supplied\,to\,combustor\,[mass\,flow \times lower\,heating\,value]}) \times 100 \atop + recirculated\,exhaust\,[mass\,flow \times lower\,heating\,value]} \tag{4}$$

φ^*: apparent equivalence ratio calculated from fuel and oxidizer without O_2 concentration in recirculating dilution gas, [-]

Acknowledgements

The author wishes to express their appreciation to the many people who have contributed to this investigation.

Author details

Takeharu Hasegawa

Address all correspondence to: takeharu@criepi.denken.or.jp

Central Research Institute of Electric Power Industry, Nagasaka, Yokosuka-Shi Kanagawa-Ken, Japan

References

[1] BP Statistical Review, "Historical Statistical Data from 1965-2010" and "BP Statistical Review of World Energy 2011", http://www.bp.com/statisticalreview/ (accessed on 13 March 2011).

[2] for example, The Energy Data and Modelling Center, The Energy Conservation Center, Japan, 2010, "2010 EDMC Handbook of Energy & Economic Statistics in Japan", ISBN:978-4-87973-365-8.

[3] Shirai,H., et al., 2007, "Proposal of high efficient system with CO_2 capture and the task on integrated coal gasification combined cycle power generation," Central Research Institute of Electric Power Industry (CRIEPI) Report No.M07003. (in Japanese)

[4] Nakao,Y., et al., 2009, "Development of CO_2 capture IGCC system -Investigation of aiming at higher efficiency in CO_2 capture IGCC system-," CRIEPI report No.M08006. (in Japanese)

[5] Hasegawa,T., et al., 2011, "Study on gas turbine combustion for highly-efficient IGCC power generation with CO_2 capture -2nd report: emission analysis of gasified-fueled gas turbines with circulating exhaust & stoichiometric combustion-," CRIEPI report No.M10005, ISBN:978-4-7983-0462-5. (in Japanese)

[6] Hasegawa,T., 2012, "Combustion Performance in a Semi-Closed Cycle Gas Turbine for IGCC Fired with CO-Rich Syngas and Oxy-Recirculated Exhaust Streams," Trans. ASME, J. Eng. Gas Turbines Power, Vol.134?, Issue *, pp.***-***, ISSN:0742-4795. (in press)

[7] Kobayashi,M., et al., 2009, "Optimization of dry desulfurization process for IGCC power generation capable of carbon dioxide capture -determination of carbon deposition boundary and examination of countermeasure-," CRIEPI report No.M09015. (in Japanese)

[8] Umemoto, S., et al., 2010, "Modeling of coal char gasification in coexistence of CO_2 and H_2O," Proceedings of The 27th Annual International Pittsburgh Coal Confer-

ence, The University of Pittsburgh, Hilton Istanbul, Istanbul, TURKEY (13 October 2010).

[9] Kidoguchi, K., et al., 2011, "Development of oxy-fuel IGCC system with CO_2 recirculation for CO_2 capture -experimental examination on effect of gasification reaction promotion by CO_2 enriched using bench scale gasifier facility-," Proceedings of the ASME Power Conference 2011 and the International Conference on Power Engineering 2011 (14 July 2011).

[10] Engineering Advancement Association of Japan, WE-NET Home Page/WE-NET report, http://www.enaa.or.jp/WE-NET/report/report_j.html (accessed on 12 March 2011).

[11] Ciferno,J.P., et al., 2010, "DOE/NETL carbon dioxide capture and storage RD&D roadmap (December 2010)," available online: http://www.netl.doe.gov/technologies/carbon_seq/refshelf/CCSRoadmap.pdf (accessed on 13 September 2011).

[12] Ciferno,J.P., et al., 2011, "DOE/NETL Advanced CO_2 Capture R&D Program: Technology Update, May 2011 Edition," available online: http://www.netl.doe.gov/ technologies/coalpower/ewr/pubs/CO_2CaptureTechUpdate051711.pdf (accessed on 1 November 2011).

[13] Bancalari,Ed., Chan,P., and Diakunchak,I.S., 2007, "Advanced hydrogen gas turbine development program," Proceedings of the ASME Turbo Expo 2007: Power for Land, Sea and Air GT2007, ASME paper GT2007-27869, Montreal, Canada, May 14-17, 2007.

[14] Todd,D.M. and Battista,R.A., 2000, "Demonstrated applicability of hydrogen fuel for gas turbines," Proceedings of the IChemE "Gasification 4 the Future" Conference, Noordwijk, the Netherlands, April 11-13, 2000.

[15] Kajitani,S., Suzuki,N., Ashizawa,M., and Hara,S., 2006, "CO2 gasification rate analysis of coal char in entrained flow coal gasifier," Fuel 85, pp.163-169.

[16] Nunokawa,M., Kobayashi,M., Shirai,H., 2008, "Halide compound removal from hot coal-derived gas with reusable sodium-based sorbent," Powder Technology, 180(1-2), pp.216-221.

[17] Akiho,H., Kobayashi,M., Nunokawa,M., Tochihara,Y., Yamaguchi,T., Ito,S., 2008, "Development of Dry Gas Cleaning System for Multiple Impurities -Proposal of low-cost mercury removal process using the reusable absorbent - ," Central Research Institute of Electric Power Industry, Report No.M07017. (in Japanese)

[18] Ozawa,Y. and Tochihara,Y., 2010, "Study of Ammonia Decomposition in Coal Derived Gas - Decomposition Characteristics over Supported Ni Catalyst -," Central Research Institute of Electric Power Industry, Report No.M09002. (in Japanese)

[19] Nunokawa,M., Kobayashi,M., Nakao,Y., Akiho,H., Ito,S., 2010, "Development of gas cleaning system for highly-efficient IGCC -Proposal for scale-up scheme of optimum

gas cleaning system based on generating efficiency analysis-," Central Research Institute of Electric Power Industry, Report No.M09016. (in Japanese)

[20] for example; Hasegawa,T., 2006, "Gas turbine combustor development for gasified fuels and environmental high-efficiency utilization of unused resources," Journal of the Japan Petroleum Institute, Vol.49, No.6, pp.281-293.

[21] Li,H., Yan,J., and Anheden,M., 2009, "Impurity inpacts on the purification process on Oxy-fuel comubtion based on CO_2 capture and storage system," Appl. Energy2009; 86, pp.202-213.

[22] Miller,J.A. and Bowman,C.T., 1989, "Mechanism and modeling of nitrogen chemistry in combustion," Prog. Energy Combust. Sci., vol.15, pp.287-338.

[23] Hasegawa,T., et al., 1998, "Study of ammonia removal from coalgasified Fuel," Combust. Flame 114, pp.246258.

[24] Lyon,R.K., 1979, "Thermal DeNOx: how it works," Hydrocarbon Processing 1979 (ISSN 0018-8190), Gulf Publishing Co., Houston, Texas, (October 1979), vol.59, No.10, 109-112.

[25] Hasegawa,T., Sato,M., Sakuno,S., and Ueda,H., 2001, "Numerical Analysis on Apprication of Selective Non Catalytic Reduction to Wakamatsu PFBC Demonstration Plant", Paper number : JPGC2001/FACT-19052, Presented at the 2001 International Joint Power Generation Conference, New Orleans, Louisiana, U.S.A., 2001-6-6.

[26] Nakata T., Sato M., and Hasegawa T., 1998, "Reaction Kinetics of Fuel NOx Formation for Gas Turbine conditions", Trans. ASME, J. Eng. Gas Turbines Power, Vol.120, No.3, pp.474-480.

[27] Hasegawa,T., et al., 1997, "Fundamental Study On Combustion Characteristics Of Gas Turbine Using Oxygen-Hydrogen -Numerical Analysis Of Formation/Destruction Characteristics Of Hydrogen, Oxygen And Radicals-," CRIEPI report No.W96007. (in Japanese)

[28] C.Y. Liu, G. Chen, N. Sipöcz, M. Assadi, X.S. Bai, 2012, "Characteristics of oxy-fuel combustion in gas turbines," Applied Energy, Volume 89, Issue 1, January 2012, pp. 387-394.

[29] Chase, Jr.M.W.; Davies, C.A.; Downey, Jr.J.R.; Frurip, D.J.; McDonald, R.A.; Syverud, A.N., 1985, "JANAF Thermodynamical Tables, 3rd Edition.," J. Phys. Chem. Reference Data, Vol.14.

[30] Kee,R.J., Rupley,F.M., and Miller,J.A., 1990, "The CHEMKIN Thermodynamic Data Base," Sandia Report, SAND 878215B.

[31] Hindmarsh,A.C., 1974, "GEAR: Ordinary differential equation system solver," Lawrence Livermore Laboratory, Univ. California, Report No. UCID30001, Rev.3.

[32] Pratt, D.T.; Bowman, B.R.; Crowe, C.T. Prediction of Nitric Oxide Formation in Turbojet Engines by PSR Analysis. AIAA paper 1971, No.71–713.

[33] Engineering Advancement Association of Japan, WE-NET Home Page/WE-NET report, http://www.enaa.or.jp/WE-NET/report/1998/japanese/gif/823.htm#823 (accessed on 12 March 2012).

Gas Turbine Combustion

Experimental Investigation of the Influence of Combustor Cooling on the Characteristics of a FLOX©-Based Micro Gas Turbine Combustor

Jan Zanger, Monz Thomas and Aigner Manfred

Additional information is available at the end of the chapter

1. Introduction

Decentralised combined heat and power generation (CHP) offers a higly efficient and sustainable way for domestic and industrial energy supply. In contrast to electrical power generation by large scale power plants in the MW-range the waste heat of the electical power generation of Micro-CHP-Units can be used by the customer without extensive grid losses. Using the combined heat and power concept overall power plant efficiencies of up to 90 % for sub MW-range CHP-units are possible [1]. In recent CHP plants conventional piston gas engines are mostly used since these systems show a good electric efficiency paired with moderate investment costs. On the other hand cycles based on micro gas turbine (MGT) systems have the potential to play an important role in decentralised power generation. In small plants for distributed power generation the flexible application of different gaseous fuels (natural gas qualities, bio fuels and low calorific gases) is an important factor. Furthermore, the national standards for exhaust gas emission levels need to be met not only at the time of installation but also after years of operation. Here, compared to piston engines MGT systems have advantages regarding fuel flexibility, maintenance costs and exhaust gas emissions [2]. This gives the possibility to avoid the installation of a cost-intensive exhaust gas treatment. Due to higher exhaust gas temperatures MGTs are more suitable for the generation of process heat and cooling. Furthermore, MGTs can be operated in a wider range of fuel gas calorific value and they are less sensitive to the fuel gas composition. Beside the advantages MGT systems need to be optimised in terms of electric efficiency which is recently at $\approx$ 30% in natural gas operation.

In order to increase fuel flexibility, electrical efficiency, product life time and reliability of micro gas turbine systems while meeting today's and future exhaust gas emission levels, further development of the combustion systems needs to be done. To meet these tasks innovative combustion concepts are needed.

Modern combustion systems for MGTs are mainly based on swirl-stabilised lean premixed concepts which promise low levels of exhaust gas pollutants. Here, the central recirculation zone induced by the swirl of the income air flow serves to stabilise the flame of the main stage generating compact flames. However, these combustion systems tend to a liability to thermo-acoustic instabilities and coherent flow structures like PVC, especially at lean premixed conditions [3, 4] which can cause high amplitude pressure oscillations. This can lead to serious damage of components in the combustion system as well as the turbo engine itself. In addition, oscillating pressure and heat release zones as well as local flame extinction, caused by coherent flow structures can have a huge impact on the production of combustion emissions. Moreover, swirl-stabilised combustion is limited in using different fuel gas compositions regarding flame flashback incidences [5] and reliable operation range. In particular, fuel gas compositions with high hydrogen fractions limit the application of swirl-stabilised combustion concepts. Studies have shown the potential of the Flameless Oxidation (FLOX©)[6] based jet-stabilised combustion concept to achieve both low exhaust gas emission levels and reduced risk towards thermo-acoustic instabilities in combination with high fuel flexibility. As discussed by Hambdi et al. [7] the general idea of this concept is also known as MILD combustion [8], colourless distributed combustion [9] or high temperature air combustion (HiTAC) [10] to name a few. The main characteristic which all those similar concepts share is the use of high temperature process air and a high dilution of the fresh gas mixture by recirculated flue gases. In particular the FLOX©-concept is characterised by non-swirled technically premixed high impulse jets penetrating a combustion chamber in a circular arrangement. These jets drive a strong inner recirculation resulting in an effective mixing process of hot exhaust gases and the fresh incoming fuel/air mixture. This enhances the flame stabilisation but also reduces the chemical reaction rates by a strong dilution. Hence, the reaction zone is stretched over a larger volume compared to swirl-stabilised combustion concepts. This volumetric reaction region exhibits an almost homogeneously distributed temperature profile inside the combustion chamber close to the adiabatic flame temperature of the global equivalence ratio Φ promising low NOx emission levels [11]. Due to the high momentum jets and therefore, the absence of low velocity zones of the income air mixture, the combustion concept has a high resistance to flashback incidents even at highly premixed conditions [12] and high hydrogen fractions [13].

Typical temperature and velocity fields of a FLOX©-based combustor are exemplarily reported by Schütz et al. [14]. Combustion stability, limits of the flameless regime as well as a comparison between experimental and numerical results obtained by Large-Eddy Simulation is reported by Duwig et al. [15]. Lückerath et al. [12] compared OH-PLIF and OH*-chemiluminescence images of a jet-stabilised burner for thermal powers up to 475kW at elevated pressure. Major species concentrations, velocity and temperature fields as well as reaction regions were reported by Lammel et al. [16] for a generic single nozzle setup using particle image velocimetry, laser raman spectroscopy as well as OH-PLIF measurements.

Beside the discussed advantages the implementation of a FLOX©-based combustion concept to a MGT system poses some challenges and tasks. First of all, the quality of the air/fuel premixture has a significant influence on the flame characteristics and emission levels. Therefore, the combustor has to be carefully designed to generate an optimised air/fuel exit profile. Moreover, since MGT systems need to have competitive prices compared to piston engines, the turbo components usually exhibit a most simple design. Therefore, the turbine blades are not cooled internally resulting in much lower turbine inlet temperature limits compared to industrial gas turbines. In order to adjust the turbine inlet temperature profile a considerable amount of cool compressor air bypasses the combustion chamber and remixes with the hot exhaust gases at the combustion chamber exit. Since the FLOX© combustion at high air numbers exhibits volumetric reaction regions, the combustion requires more room compared to swirl-stabilised combustion concepts. The low production of harmful emissions as well as the high flame stability of the FLOX©-regime would be negatively influenced by the injection of cold dilution air into extensively expanded volumetric reaction zones. Therefore, combustion system design parameters like length of the combustion chamber and position, shape and pattern of dilution holes have to be considered for a final combustor development.

In order to improve flame stability, lifetime, operating range and exhaust gas emissions of the combustion system in a commercial Turbec T100 MGT, a FLOX©-based jet-stabilised combustor was designed for natural gas utilising the advantages of this concept. The recent work covers an experimental study of two combustor configurations differing in the combustor front plate cooling. The paper presents the influence of the combustor cooling air on flame characteristics, lean blow off (LBO) limits and exhaust gas emissions. Flame characteristics are analysed, using measurements of the OH*-chemiluminescence (OH*-CL) signal at selected power loads and air numbers. From these images the height above burner, the dispersion of OH*-CL signal and its homogeneity are derived. These quantaties are discussed with respect to the effects of cooling air, thermal power load and air number on the combustor performance.

2. Experimental setup

2.1. Combustor design

In Figure 1 and 2 both FLOX-based combustor configurations are shown which were used in this study. The reference combustor without any combustor front plate cooling is displayed in Figure 1. In order to increase the combustor lifetime an additional impingement front plate cooling is implemented into the second configuration to decrease the temperature at the location of the highest thermal loads. Both combustor designs consist of 20 fuel and air nozzles in a circular arrangement. Natural gas is injected concentrically into the air nozzles, which are oriented co-axially with respect to the combustion chamber. Both air and fuel gas flow are injected without any swirl. Partial premixing of fuel and air is achieved by the special design of the injecting system. Vortical structures generated by the air injection system provide a macroscopic mixing of fuel and air. Turbulence produced by the fuel injection system provides mixing on a microscopic scale. For the impingement cooling system a small part of the overall air mass flow is fed into a cooling air plenum and passes a perforated plate. This perforated plate produces small cooling air jets impinging the combustor front plate from the back side. After the cooling air has hit the combustor front plate it is led radially through channels and injected into the combustion chamber through 20 holes. The cooling air holes are positioned in between the FLOX© combustor air nozzles on a slightly larger diameter. In addition to the impingement cooling the front plate of the cooled design is coated with a zirconium oxide thermal protection layer.

2.2. Test rig description

In order to run parametric studies of the combustion performance independent from the MGT load point limitations, the combustor was implemented into an atmospheric test rig.

Experimental Investigation of the Influence of Combustor Cooling on the Characteristics of a FLOX®-
Based Micro Gas Turbine Combustor

145

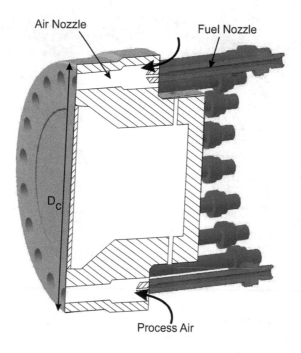

Figure 1. Uncooled Combustor.

The experimental setup shown in Figure 3 comprises the air and fuel supply, the combustor and an optically accessible, hexagonal combustion chamber comprising six quartz glass windows. The hexagonal cross-section was chosen as a trade-off between good optical accessibility and the analogy to the circular cross-section of the original MGT liner. A circular fuel plenum which is situated under the air plenum but not shown in Figure 3 ensures an equal supply of all 20 fuel gas nozzles. In order to emulate the combustor inlet conditions of the MGT the air can be preheated electrically up to 925K by five 15kW "Leister" air heater units. The complete air supply system is decoupled acoustically from the test bench by a perforated plate located at the air inlet. When entering the air plenum the air flow is directed via a baffle in a way that a 180°deflection at the combustor inlet of the original MGT is reproduced. This was found to be essential to generate a specific premixing profile in the combustor nozzle and hence is important for flame characteristics and flame stabilisation. After passing the baffle a small part of the air flow enters the cooling plenum and the major part is fed into the air nozzles of the combustor, where the premixing with the fuel takes place. In this study the combustor is operated with natural gas (LHV = 47.01 $\frac{MJ}{kg}$, AFR$_{stoech}$ = 16.2). On the top of the combustion chamber an exhaust gas duct is flanged.

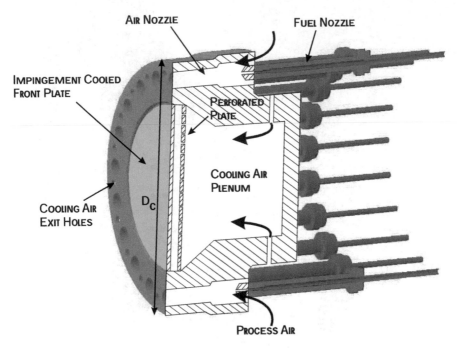

Figure 2. Impingement Cooled Combustor.

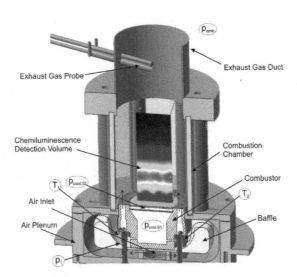

Figure 3. Atmospheric Combustor Test Rig

2.3. Instrumentation

To analyse all relevant process parameters inside the system the test rig is equipped with a detailed instrumentation. The data acquisition at a frequency of 2 Hz is realised by "Delphin" modules. For temperature measurements a total number of 27 thermocouples (type N, precision class 2) are installed. The arithmetic average of the temperature T_1 and T_2 which are situated in the air flow at the combustor inlet defines the combustor preheat temperature T_V with an uncertainty of $\pm 0.85\%$ of the actual value. The rig furthermore comprises 2 total and 13 static pressure transducers read out by pressure scanners "Netscanner Model 9116" and "Model 9032" by Esterline Pressure Systems. All pressures can be optained with a manufacturer's accuracy of ± 4 mbar. Combustor pressure loss is determined by the static pressure p_1 measured short before the combustor inlet (see figure 3) and the ambient pressure. Applying a suitable calibration, the mass flow through the cooling system is calculated as a function of the pressure loss between the static pressures $p_{cool,01}$ and $p_{cool,02}$ situated in front of and behind the perforated plate. The fuel mass flow is controlled by a "Bronkhorst Cori-Flow" coriolis mass flow controller with a manufacturer's accuracy of $\pm 0.5\%$ of the actual value and the air mass flow is regulated by a "Bronkhorst EL-Flow" thermal mass flow controller with a manufacturer's accuracy of $\pm 0.8\%$. As indicated in figure 3 a radially traversable suck-up exhaust gas probe is mounted inside the exhaust gas duct. The probe is equipped with a coaxial air cooling keeping the probe tip at a constant temperature of 120°C to achieve a sufficient quenching of the measured exhaust gas. This ensures defined measuring conditions. The sucked-up exhaust gases are directed via heated hoses to an "ABB" exhaust gas analysing system. The flue gas species O_2, CO, CO_2, NO, NO_2 and unburned hydrocarbons (UHC) are measured by a magnetomechanical analyser "Magnos106", a infrared analyser "Uras14", a UV photometer "Limas11 HW" and a flame ionisation detector "MultiFID14". The species O_2, CO and CO_2 are measured in a dry environment, whereas all other species are detected in wet conditions. The measurements of the species shown in this study have manufacturer's accuracies as indicated in Table 1.

	CO [ppm]	NO_x [ppm]	UHC [ppm]	O_2 [Vol-%]
Range 1	0-8	0-24	0-9	0-25
Accuracy 1	0.1	0.5	0.1	0.25
Range 2	8-80	24-238	9-90	
Accuracy 2	1	5	1	

Table 1. Ranges and Corresponding Accuracies of the Measured Exhaust Gas Species.

OH* chemiluminescence measurements were used to study the shape, location and homogeneity of the heat release zone. The electronically excited OH* radical is formed by chemical reactions in the reaction zone, predominately via $CH + O_2 \rightarrow CO + OH^*$ [17]. Since its lifetime is very short, the emitted OH*-CL signal originates only from within the reaction region. Therefore, the OH*-CL signal is a very good marker for the location and dimension of the reaction zone. However, this technique is a line-of-sight method giving only spatially integrated information in the combustion chamber depth. The OH*-CL emissions were imaged using a "LaVision FlameStar 2" intensified CCD camera in combination with a "Halle" 64mm, f/2 UV lens and a UV interference filter ($\lambda = 312 \pm 20$nm). All OH*-CL data was time-averaged over a time series of 200 instantaneous images acquired with a repetition rate of 3.6Hz. Due to the substantial difference of the OH*-CL signal over the complete load point range the gate width was varied between 18 and 600 μs at maximum gain factor.

As indicated in Figure 4 the detection volume covers four air nozzles on each side of the combustor. Due to assembly restrictions the nozzles on opposing sides are arranged with a small misalignment with respect to the line of sight.

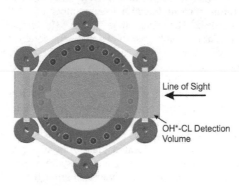

Line of Sight

OH*-CL Detection Volume

Figure 4. Line of Sight of the OH*-CL measurements

3. Experimental results

The measurements shown in this section were carried out at steady-state combustion conditions, whereupon every single load point is time-averaged over 5 min at an acquisition rate of 2Hz. In order to analyse the operating range of the combustor configurations the overall air number $\lambda_{overall}$ and the thermal power were varied at a constant combustor inlet air preheat temperature T_v. In this study the normalised thermal power $Q_{th,n}$ with respect to MGT full load conditions is presented. The overall air number is defined as the reciprocal of the fuel equivalence ratio Φ calculated with the overall air and fuel mass flows. All mass flows of the MGT load points were scaled to atmospheric conditions keeping the velocity fields constant. Table 2 shows the matrix of the parametric study.

Parameter	Range	Unit
$Q_{th,n}$	35 - 100	%
$\lambda_{overall}$	1.8 - LBO	-
T_v	580	°C

Table 2. Ranges of the Measuring Matrix.

3.1. Operating range

Figure 5 visualises the operating range of the uncooled combustor configuration displaying the normalised axial jet velocity u_{nozzle} at the exit of a single air nozzle as a function of $Q_{th,n}$. U_{nozzle} is defined as

$$u_{nozzle} = \frac{\dot{m}_{nozzle} \cdot R_{mix} \cdot T_{c,in}}{p_{c,in} \cdot A_{nozzle}} / u_{nozzle,max} \qquad (1)$$

with the combustor inlet temperature $T_{c,in}$, the static combustor inlet pressure $p_{c,in}$, the planar-averaged specific gas constant of the air/fuel mixture R_{mix}, the nozzle cross-section

area A_{nozzle} and the averaged mass flow of a single nozzle $\dot{m}_{nozzle}$. The whole measuring field
of the operating range was scanned by adjusting a constant $Q_{th,n}$ and increasing the $\lambda_{overall}$
up to lean blow-off (LBO) conditions. In this study the lean blow-off is defined as the point
where the flame actually extinguishes. The filled points symbolise the feasible operating
points whereas the blank points emblematise the points at which LBO occured. All illustrated
load points represent an individual measurement which means that no information about the
reproducibility of the LBO limit can be given. Regarding the points around LBO it is visible
that the feasible $\lambda_{overall}$ slopes with increasing thermal power. The LBO at $Q_{th,n}$ = 35% is with
$\lambda_{overall}$ = 3.26 substantially higher compared to $Q_{th,n}$ = 100% with $\lambda_{overall}$ = 3.01. Moreover,
figure 5 indicates that higher jet velocities can be realised with increasing $Q_{th,n}$ before LBO
occurs. The axial jet velocity u_{nozzle} at LBO significantly varries from 58% at $Q_{th,n}$ = 35% to
approximately 140% at $Q_{th,n}$ = 100%. This effect was also observed and discussed by Vaz et
al. [18] for a similar system.

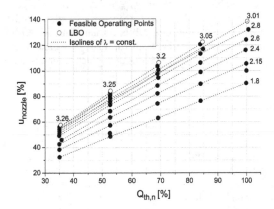

Figure 5. Operating Range of Uncooled Combustor Configuration.

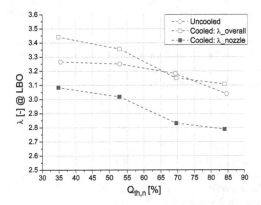

Figure 6. Air Number at LBO of the Cooled and Uncooled Combustor Configurations.

In Figure 6 the comparison between the LBO conditions of the uncooled combustor and the cooled configuration is presented as a function of $Q_{th,n}$. Due to mass flow limitations the LBO of the cooled design could only be measured for $Q_{th}/Q_{th,max} \leq 85\%$, therefore, only this range is visualised. For the cooled design the overall air number $\lambda_{overall}$ of the combustor is shown as well as the local air number λ_{nozzle} of a single nozzle which is reduced by the cooling air mass flow. However, for the uncooled design the local and the overall air numbers are equal due to the lack of cooling air. The $\lambda_{overall}$ at LBO of the cooled configuration follows the general sloping trend of the uncooled design for increasing $Q_{th,n}$ but exhibits a steeper gradient and with $\lambda_{overall}$ = 3.44 at $Q_{th,n}$ = 35% higher values at low thermal loads. This means using a cooled combustor configuration that for $Q_{th}/Q_{th,max} \leq$ 52% more overall air can be fed into the combustion chamber before LBO occurs. This behaviour is advantageous if the amount of combustion air is to be maximised. However, for $Q_{th}/Q_{th,max} \geq 70\%$ the difference between both designs is negligible. Furthermore, the graph of the cooled design exhibits a kink at $Q_{th,n}$ = 70%. This behaviour is reproducible but in order to give a conclusive explanation further investigation need to be carried out. Regarding λ_{nozzle} the cooled configuration exhibits a much lower LBO limit for the whole operating range, whereas the difference between the designs increases with rising thermal power. This indicates that the fraction of cooling air, which interacts with the combustion process, changes with $Q_{th,n}$.

3.2. Exhaust gas emissions

The carbon monoxide (CO) exhaust gas emission profiles of the uncooled combustor configuration are shown in figure 7 as a function of $\lambda_{overall}$ for all thermal powers, whereas CO is normalised to 15% oxygen concentration. All curves exhibit a similar U-shape with a distinct minimum. Regarding the graph at $Q_{th,n}$ = 100% the CO concentrations decrease from 50 ppm at $\lambda_{overall}$ = 1.8 to 24 ppm at $\lambda_{overall}$ = 2.42 followed by a rise up to 58 ppm at $\lambda_{overall}$ = 2.8. In the left hand branch of the CO curves the trend of the measured concentrations follow the trend of the equilibrium conditions which decrease for higher air numbers [19]. The right hand branch, however, is dominated by non-equilibrium effects [20]. Here, the residence time of the flue gases inside the combustion chamber before reaching the exhaust gas probe is insufficiently long to achieve equilibrium state resulting in higher measured CO concentrations. Moreover, due to the expansion of the reaction zones at LBO-near overall air numbers, decribed in section 3.3, the exhaust gas probe is for these conditions located inside the reaction zone resulting in incomplete combustion process at the position of measurement. Regarding the CO profiles at different thermal powers a destinct layered arrangement is observable showing higher CO concentrations for increasing $Q_{th,n}$. Here again, the influence of the residence time is visible since at constant $\lambda_{overall}$ higher thermal powers exhibit higher fuel and air mass flows reducing the overall residence time.

In figure 8 the comparison between the CO profiles of the uncooled combustor and the cooled configuration is shown as a function of λ_{nozzle}. For clarity reasons only the profiles for $Q_{th,n}$ = 53% and 100% are visualised exemplarily but it should be mentioned that the trend of the displayed curves applies to the other thermal powers as well. The CO profiles of both combustor designs show similar U-shaped trends. But it is clearly visible that the profiles of the cooled configuration are shifted horizontally to lower air numbers. The magnitude of the shift is thereby dependent on thermal power. For $Q_{th,n}$ = 100% the shift seems to be approximately constant for all λ_{nozzle}, however, for $Q_{th,n}$ = 53% the CO profiles fit well at

Experimental Investigation of the Influence of Combustor Cooling on the Characteristics of a FLOX®-
Based Micro Gas Turbine Combustor

151

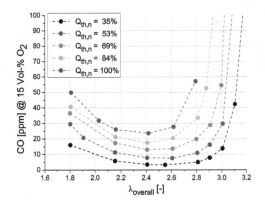

Figure 7. CO Emissions of the Uncooled Combustor Configuration.

low air numbers but differ at high λ_{nozzle}. Due to a high sensitivity to the cooling air mass flow's error the maximum uncertainty of λ_{nozzle} after propagation of error is approximately ±5% for the cooled configuration. However, the uncooled design exhibits an uncertainty better than 1% due to the lack of cooling air. Nevertheless, since the observed shift of the CO profiles shows a distinct systematic trend, it is believed to represent a physical effect.

When comparing the CO emissions of both combustor designs with respect to $\lambda_{overall}$, see figure 9, the difference between both configurations is even more pronounced but shifted to the opposite direction compared to the display as a function of λ_{nozzle}. In the case that the CO profiles of the cooled configuration were equal to the uncooled design for $\lambda_{overall}$ this would suggest that all cooling air participates in the reaction process. On the other hand equal CO profiles for λ_{nozzle} would mean that no cooling air enters the reaction region. Therefore, the observed behaviour indicates that only a part of the cooling air participates in the primary reaction zone.

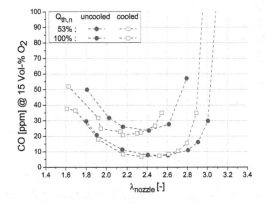

Figure 8. CO Emissions of the Cooled and Uncooled Combustor Designs as a Function of λ_{nozzle}.

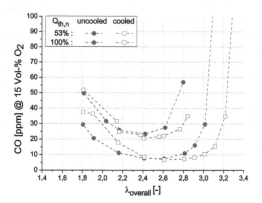

Figure 9. CO Emissions of the Cooled and Uncooled Combustor Designs as a Function of $\lambda_{overall}$.

In order to approximate the amount of cooling air which interacts with the reaction region, the λ_{nozzle} of the cooled configuration is modified in a way that the CO profiles of both combustor designs fit well as shown in figure 10. This new air number is called $\lambda_{nozzle}(\mathrm{mod})$. For the cooled design it is approximately $\lambda_{nozzle}(\mathrm{mod}) \approx \lambda_{nozzle} + 0.1 \neq const.$ and for the uncooled configuration it equals λ_{nozzle}. This modified air number, which is based on the CO emissions, is an auxiliary quantity for comparing both combustor configurations. Since the visualisation as a function of $\lambda_{nozzle}(\mathrm{mod})$ neutralises the shift of air number between both designs, the comparison of the magnitudes of certain quantities are facilitated. Therefore, the modified air number is used in the following sections to characterise the difference of both combustor designs.

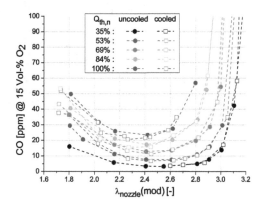

Figure 10. CO Emissions of the Cooled and Uncooled Combustor Configurations as a Function of the Modified Air Number $\lambda_{nozzle}(\mathrm{mod})$.

The nitrogen oxide (NOx) emissions at 15% O_2 of both combustor designs are visualised in figure 11 as a function of $\lambda_{nozzle}(\mathrm{mod})$ for $Q_{th,n} = 53\%$ and 100%. All NOx profiles show a similar exponential decreasing characteristics and similar magnitudes for rising modified nozzle air numbers. For $Q_{th,n}$ = 100% the NOx emissions reduce from 18ppm at λ_{nozzle}(mod) = 1.75 down to 2 ppm at λ_{nozzle}(mod) = 2.8 for both combustors. Since the measurements were conducted at lean atmospheric conditions using natural gas, the major NOx formation mechanism is the thermal dominated Zeldovich mechanism. Therefore, the trend of the NOx profiles reflects the exponential decrease of the thermal NOx formation with falling flame temperature and rising air number, respectively. With respect to the profiles at different thermal powers, a very low dependence on $Q_{th,n}$ can be observed. For the uncooled design the curves of all thermal powers match very well, however, for the cooled configuration the NOx profiles at $Q_{th,n}$< 100% exhibit slightly higher magnitudes at low λ_{nozzle}(mod) compared to full load conditions. Regarding the fact that the NOx emissions of both combustor designs fit very well as a function of λ_{nozzle}(mod), the significance of this modified nozzle air number for comparing the configurations is backed up. This affirms the assumption that only a part of the cooling air interacts with the reaction region.

The emission levels of unburned hydrocarbons (UHC), not shown in this section, are under the detection threshold for both combustors and all thermal powers apart from air numbers close to the LBO. Here, a rapid increase of UHCs is detectable due to incomplete combustion close to the blow-off limit.

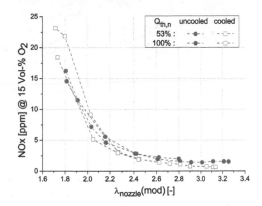

Figure 11. NOx Emissions of the Cooled and Uncooled Combustor Configurations as a Function of the Modified Air Number λ_{nozzle}(mod).

3.3. Flame shape and location

The flame characteristics regarding shape, location and homogeneity are discussed for both combustor designs in this section using time-averaged OH*-CL images. Figure 12 shows a series of OH*-CL images for the uncooled combustor configuration at $Q_{th,n}$ = 69% as a function of $\lambda_{overall}$. All images are scaled between zero and their maximum signal intensity. The corresponding scaling factors are indicated in the upper left corner. The points of origin of abscissa and ordinate mark the centre axis of the combustor and the combustor front plate, respectively. The dimensions are normalised with respect to the combustor diameter. Moreover, the azimuthal positions of the air nozzles located at the window in front of the camera are indicated at the bottom. Regarding the images at low overall air numbers discrete reaction zones around the entering fresh gas jets can be observed. Here, the reaction zones are characterised by compact shape and distinctly separated flames. For air numbers between 2.2 and 2.4 the flame length stays approximately constant, whereas the height above burner (HAB) increases with rising $\lambda_{overall}$. Furthermore, for $2.2 \leq \lambda_{overall} \leq 2.8$ the reaction zones continuously merge into each other in horizontal direction evolving from separated flames to a single reaction region. For air numbers above 2.6 the reaction zone spreads in all spatial directions. Simultaneously, the HAB declines whereas the length of the reaction zone grows substantially. At $\lambda_{overall} = 3.1$, which is the last operating point before LBO occured, the reaction zone is distributed over almost the whole combustion chamber volume. Moreover, regarding the scaling factors of the images the signal intensity decreases significantly with rising air numbers leading to a blueish visible flame of very low luminosity for $\lambda_{overall} \geq 2.8$.

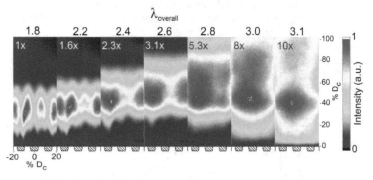

Figure 12. Time-averaged OH*-CL Images of the Uncooled Combustor at $Q_{th,n} = 69\%$ as a Function of $\lambda_{overall}$.

In order to quantify and compare the phenomena observed in the time-averaged OH*-CL images, characteristic parameters are derived from the data. For the subsequent analysis only signals above 50% of the image maximum intensity are taken into account. This definition has been found as the most appropriate method to cover the image pixels corresponding visually to the flame. In the following these pixels are called reaction or flame region. With this definition the HAB equals the axial distance between the combustor front plate and the horizontally averaged lower flame boundary. The Dispersion of the OH*-signal is defined as the area of the flame region divided by the overall area of OH*-measurement. Therefore, the dispersion is a marker for the relative reaction volume, but not for its homogeneity. The horizontal distribution of the reaction regions is evaluated by the Relative Horizontal Inhomogeneity dI_{Flame}/dx. This parameter is calculated by vertically averaging the horizontal spatial intensity gradients inside the reaction regions and normalising it to the flame average intensity of each image. This definition ensures that dI_{Flame}/dx is comparable for different signal intensity levels and flame shapes. The Relative Horizontal Inhomogeneity is a marker for the discreetness of the flames in horizontal direction declining for merging reaction regions.

Figure 13 visualises the Dispersion of OH*-signal of the uncooled combustor design for all thermal powers as a function of $\lambda_{overall}$. Regarding $Q_{th,n} = 100\%$ the dispersion stays approximately constant at 25% between $\lambda_{overall} = 1.8$ and 2.15. For $\lambda_{overall} \geq 2.4$ the magnitude increases significantly up to 73%. This means that with rising Dispersion of OH*-signal the occupied volume of the reaction region inside the combustion chamber increases substantially until the flame is distributed over almost the whole volume. This indicates a decrease of the Damköhler number which means that the chemical time scale increases in relation to the fluid dynamic time scale [21, 22]. An important influence factor for this behaviour is the exhaust gas recirculation rate which is enhanced by higher jet velocities and higher air numbers [11], respectively. Increasing recirculation serves to enhance the dilution of the fresh gas jets by hot flue gases [23] reducing the chemical reaction rates [24]. Simultaneously, rising jet velocity decreases the fluid dynamic time scale. Thus, both effects lead to a declining Damköhler number. On the other hand by increasing the air number the premixing quality of the air/fuel jets is altered as well. Since these quantities cannot be separated in the recent study, the major influence factor cannot be determined.

The general trend of the Dispersion of OH*-signal of the uncooled combustor design is similar for all thermal powers. However, the magnitude of the region of constant dispersion at low air numbers decreases for declining power from 25% at $Q_{th,n} = 100\%$ down to 14% at $Q_{th,n} = 35\%$. The point of ascending dispersion is located between $\lambda_{overall} = 2.4$ and 2.6 for all thermal powers but no distinct shift of this point can be observed for the uncooled configuration. In terms of reducing temperature peaks and simultaneously reducing NOx emissions, the rapid rise of the Dispersion of OH*-signal is desirable. However, for the application inside a MGT combustion system the available combustion chamber length limits the feasible volumetric expansion of the reaction region as discussed in section 1. At LBO-near conditions the flame region is expanded substantially so that the reaction process even continue after leaving the exhaust gas duct of the atmospheric test rig. Since operating pressure as well as wall heat loss also have a distinct effect on the flame characteristics, a final selection of the maximum feasible operating point can only be made on the base of a measurement campaign in the MGT test rig.

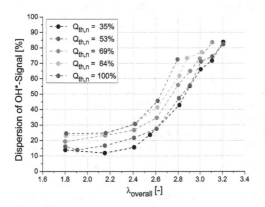

Figure 13. Dispersion of OH*-signal for the Uncooled Combustor Configuration.

Figure 14 presents the Dispersion of OH*-signal of the cooled combustor configuration for all thermal powers as a function of $\lambda_{overall}$. In contrast to the uncooled design, shown in figure 13, the cooled configuration exhibits with a range between 15% and 20% a similar magnitude of the region of constant dispersion at low air numbers for all thermal powers. Moreover, the point of ascending dispersion shifts with decreasing thermal power towards higher overall air numbers. This means that for a constant $\lambda_{overall} \geq 2.6$ higher thermal powers show a higher dispersion of the OH*-signal occupying a significantly larger combustion chamber volume.

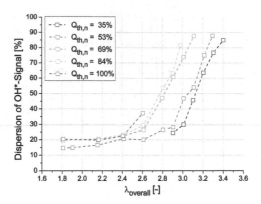

Figure 14. Dispersion of OH*-signal for the Cooled Combustor Configuration.

The height above burner of the cooled combustor design is visualised in figure 15 for all thermal powers as a function of $\lambda_{overall}$. This quantity describes the axial distance between the lower flame boundary and the combustor front plate. In the following graphs the HAB is presented normalised with respect to the combustor radius. Regarding the profile at $Q_{th,n}$ = 69% the HAB rises from 42% at $\lambda_{overall}$ = 1.8 up to 68% at $\lambda_{overall}$ = 2.6. At this air number the flame reaches its maximum lift-off height for $Q_{th,n}$ = 69%. With increasing air numbers beyond this point the HAB declines significantly exhibiting at LBO-near conditions with 23% its lowest magnitude. All HAB profiles at different thermal powers show a similar trend as well as a similar maximum lift-off height. However, for decreasing thermal power the profiles are shifted to higher overall air numbers exhibiting lower magnitudes at low air numbers. As described above, the recirculation rate intensifies at rising thermal power. Owing to higher jet velocities and enhanced fresh gas dilution by recirculated flue gases, the lift-off height increases with rising thermal power at low overall air numbers. Furthermore, the maxima of the HAB profiles are directly related to the points of ascending Dispersion of OH*-signal in figure 14. Due to the shift from discrete reaction zones into a volumetric combustion the reaction region expands to all spatial directions explaining the descent of the HAB. For high $\lambda_{overall}$ this effect of expanding reaction regions outbalances the increase of lift-off height caused by higher jet velocities at rising thermal powers which dominates at low air numbers.

The HAB profiles of both combustor designs are visualised in figure 16 as a function of $\lambda_{nozzle}(mod)$ for $Q_{th,n}$ = 53% and 100%. The graph shows that the profiles of the presented thermal powers fit very well for both combustor configurations. It should be mentioned that this behaviour is consistent with all measured thermal powers. This behaviour demonstrates that the HAB is not affected by the cooling air when both combustors are compared as a function of $\lambda_{nozzle}(mod)$. However, $\lambda_{nozzle}(mod)$ is based on a shift of λ_{nozzle} to higher air numbers for the cooled combustor configuration. Therefore, identical profiles of the cooled and uncooled designs illustrate that the cooling air shifts the flame properties to lower nozzle air numbers compared to the uncooled design. In addition, this indicates that the chosen definition of the modified nozzle air number $\lambda_{nozzle}(mod)$, which is based on the shift of the CO profiles, is in good agreement with the flame characteristics as well. This affirms the assumption that only a part of the cooling air participates in the reaction process.

Experimental Investigation of the Influence of Combustor Cooling on the Characteristics of a FLOX®-Based Micro Gas Turbine Combustor

157

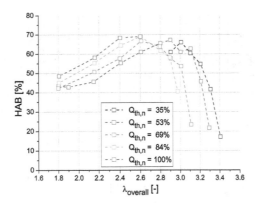

Figure 15. HAB for the Cooled Combustor Configuration as a Function of $\lambda_{overall}$.

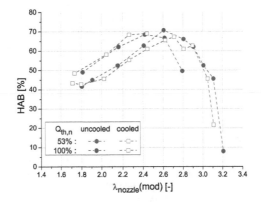

Figure 16. HAB for the Cooled and Uncooled Combustor Designs as a Function of λ_{nozzle}(mod).

Figure 17 visualises the Relative Horizontal Inhomogeneity dI_{Flame}/dx for the uncooled combustor configuration for all thermal powers as a function of $\lambda_{overall}$. This quantity describes the discreetness of the reaction zones in horizontal direction. The profile at $Q_{th,n}$ = 100% exhibits an inhomogeneity of 1.4 %/dPixel at $\lambda_{overall}$ = 1.8 converging exponentially with increasing air numbers towards a lower threshold of approximately 0.5 %/dPixel which is reached at $\lambda_{overall} \geq 2.6$. This means that for lower air numbers separated flames around the penetrating fresh gas jets exist, which merge together for rising air numbers resulting in a horizontally distributed reaction region for high $\lambda_{overall}$. The profiles of all thermal powers exhibit a lower threshold of the same magnitude which is reached at LBO-near conditions. The general trend of the profiles is similar for $53\% \leq Q_{th,n} \leq 100\%$. Only the profile at $Q_{th,n}$ = 35% differs from the exponential declining trend. However, at low overall air numbers the profiles' magnitudes are staggered in thermal power reaching higher values for low $Q_{th,n}$. This signifies that higher thermal powers exhibit better horizontally distributed flames which is due to an enhanced recirculation rate, better premixing quality at high flow rates as well as an enhanced interjet mixing rate.

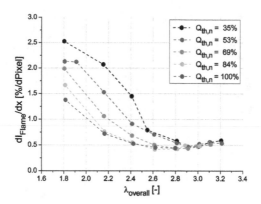

Figure 17. Relative Horizontal Inhomogeneity for the Uncooled Combustor Design as a Function of $\lambda_{overall}$.

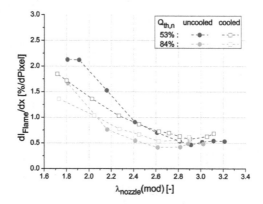

Figure 18. Relative Horizontal Inhomogeneity for the Cooled and Uncooled Combustor Designs as a Function of λ_{nozzle}(mod).

The comparison of dI_{Flame}/dx between the uncooled and the cooled combustor configuration is presented in figure 18 as a function of λ_{nozzle}(mod) for $Q_{th,n}$ = 53% and 84%. These thermal powers are chosen exemplarily since the occuring effects are well pronounced for these profiles but it should be mentioned that the discussed behaviour applies to all thermal powers. For rising λ_{nozzle}(mod) the curves of both designs show a similar declining trend of dI_{Flame}/dx at all $Q_{th,n}$. For higher air numbers the profiles of the cooled and uncooled configurations fit very well in magnitude at a constant thermal power converging to a similar lower threshold of approximately 0.5 %/dPixel. However, at low air numbers the uncooled design distinctly exhibits higher magnitudes of the Relative Horizontal Inhomogeneity. The difference of both designs decreases with rising thermal power as well as rising air number. This means that the difference in dI_{Flame}/dx decreases with increasing axial jet velocity u_{nozzle}. As shown in figure 2 the cooling air penetrates the combustion chamber through small exit holes located between the air nozzles. Since the split between cooling air and process air is almost constant over the whole operating range, the cooling jet velocity scales with the axial jet velocity u_{nozzle} and air number, respectively. Figure 18 indicates that for lower jet velocities the penetrating cooling air serves to broaden the discreet reaction regions horizontally and therefore to homogenise their distribution. However, the homogenising of the flame regions for the cooled combustor configuration seems to have no effect on exhaust gas emissions at low air numbers as shown in figure 10 and 11.

4. Conclusion

A FLOX©-based micro gas turbine combustor was introduced. The presented experimental study compared an impingement cooled combustor configuration to an uncooled design for natural gas. The influence of selected operating conditions on shape, location and homogeneity of the reaction zones was analysed under atmospheric conditions using time-averaged OH*-chemiluminescence images. Furthermore, the dependencies of jet velocity and combustor front plate cooling on LBO limits were discussed. Exhaust gas emissions were presented and a definition of a modified nozzle air number was derived from comparing the CO profiles of the cooled and uncooled design. With the help of this parameter the differences of both designs were analysed.

Regarding the parameters derived from OH*-chemiluminescence images a distinct increase of Dispersion of OH*-signal was observed for rising air numbers leading to a volumetric reaction region at LBO-near conditions. Simultaneously, the detachedness and the horizontal inhomogeneity of the reaction regions reduced substantially. The influence of the cooling air was observed to generate a shift of all emission and flame profiles to lower nozzle air numbers. However, it was discussed that only a part of the cooling air interacts with the reaction region whereas the rest of the cooling air passes the combustion chamber without participating in the combustion process. Moreover, it was shown that with the cooled combustor design higher overall air numbers can be realised at low thermal powers which is advantageous if the amount of combustion air is to be maximised.

Acknowledgements

The financial support of this work by the EnBW Energie Baden-Württemberg AG and the German Federal Ministry of Economics and Technology is gratefully acknowledged. Furthermore, the authors would like to thank Marco Graf for his support.

Author details

Zanger Jan*, Monz Thomas and Aigner Manfred

* Address all correspondence to: jan.zanger@dlr.de

German Aerospace Center, Institute of Combustion Technology, Stuttgart, Germany

References

[1] Bhatt M S (2001) Mapping of General Combined Heat and Power Systems, Energy Conversion and Management 42, pp. 115-4.

[2] Pilavachi P A (2000) Power Generation with Gas Turbine Systems and Combined Heat and Power, Applied Thermal Engineering 20, pp. 1421-1429.

[3] Fritsche D, Füri M, Boulouchos K (2007) An Experimental Investigation of Thermoacoustic Instabilities in a Premixed Swirl-stabilized Flame, Combustion and Flame 151, pp. 29-36.

[4] Huang Y, Yang V (2009) Dynamics and Stability of Lean-premixed Swirl-stabilized Combustion, Progress in Energy and Combustion Science 35, pp. 293-364.

[5] Nauert A, Peterson P, Linne M, Dreizler A (2007) Experimental Analysis of Flashback in Lean Premixed Swirling Flames: Conditions Close to Flashback, Exp Fluids 43, pp. 89-100.

[6] Wünning, J.A. and Wünning, J.G.(1997) Flameless Oxidation to Reduce Thermal NO-Formation, Prog. Energy ComlnaL Sci.23, pp. 81-94.

[7] Hamdi M, Benticha H, Sassi M (2012) Fundamentals and Simulation of MILD Combustion, Thermal Power Plants, Dr. Mohammad Rasul (Ed.), ISBN: 978-953-307-952-3, InTech. pp. 43-64.

[8] Weber R, Smarta J P, vd Kamp W (2005) On the (MILD) Combustion of Gaseous, Liquid and Solid Fuels in High Temperature Preheated Air, Proceedings of the Combustion Institute, Vol. 30, pp. 2623-2629

[9] Arghode V K, Gupta A K (2010) Effect of Flow Field for Colorless Distributed Combustion (CDC) for Gas Turbine Combustion, Applied Energy 87(5), pp. 1631-1640.

[10] Tsuji H, Gupta A K, Hasegawa T, Katsuki M, Kishimoto K, Morita M (2003) High Temperature Air Combustion, CRC Press, Florida.

[11] Li G, Gutmark E J, Stankovic D, Overman N, Cornwell M, Fuchs L, Vladimir M (2006) Experimental Study of Flameless Combustion in Gas Turbine Combustors, Proceedings of 44th AIAA Aerospace Sciences Meeting and Exhibit, AIAA 2006-546.

[12] Lückerath R, Meier W, Aigner M (2007) FLOX© Combustion at High Pressure with Different Fuel Compositions, Proceedings of GT2007 ASME Turbo Expo 2007, GT2007-27337.

[13] Lammel O, Schütz H, Schmitz G, Lückerath R, Stöhr M, Noll B, Aigner M (2010) FLOX© Combustion at High Power Density and High Flame Temperatures, Proceedings of GT2010 ASME Turbo Expo 2010, GT2010-23385.

[14] Schütz H, Lückerath R, Kretschmer T, Noll B, Aigner M (2006) Analysis of the Pollutant Formation in the FLOX© Combustion, Proceedings of GT2006 ASME Turbo Expo 2006, GT2006-91041.

[15] Duwig C, Stankovic D, Fuchs L, Li G, Gutmark E (2008) Experimental and Numerical Study of Flameless Combustion in a Model Gas Turbine Combustor, Combustion Science and Technology 180, pp. 279-295.

[16] Lammel O, Stöhr M, Kutne P, Dem C, Meier W, Aigner M (2011) Experimental Analysis of Confined Jet Flames by Laser Measurement Techniques, Proceedings of GT2011 ASME Turbo Expo 2011, GT2011-45111.

[17] Dandy D S, Vosen S R (1992) Numerical and Experimental Studies of Hydroxyl Radical Chemiluminescence in Methane-Air Flames, Combustion Science and Technology 82, pp. 131-150.

[18] Vaz D C, Buiktenen J P, Borges A R J, Spliethoff H (2004) On the Stability Range of a Cylindrical Combustor for Operation in the FLOX Regime, Proceedings of GT2004 ASME Turbo Expo 2004, GT2004-53790.

[19] Lefevbre A H, Ballal D R (2010) Gas Turbine Combustion: Alternative Fuels and Emissions – 3rd ed., CRC Press, Boca Raton.

[20] Joos F (2006) Technische Verbrennung, Springer, Berlin.

[21] Borghi R (1985) On the Structure and Morphology of Turbulent Premixed Flames, Recent Advances in Aerospace Sciences, Plenum Press, New Yorck.

[22] Sadanandan R, Lückerath R, Meier W, Wahl C (2011) Flame Characteristics and Emissions in Flameless Combustion Under Gas Turbine Relevant Conditions, Journal of Propulsion and Power 27(5), pp. 970-980.

[23] Levy Y, Rao G A, Sherbaum V (2007) Preliminary Analysis of a New Methodology for Flameless Combustion in Gas Turbine Combustors, Proceedings of GT2007 ASME Turbo Expo 2007, GT2007-27766.

[24] Li P, Mi J, Dally B B, Wang F, Wang L, Liu Z, Chen S, Zheng C (2010) Progress and recent trend in MILD combustion, Science China Technological Sciences 54(2), pp. 255-269.

Chapter 7

Review of the New Combustion Technologies in Modern Gas Turbines

M. Khosravy el Hossaini

Additional information is available at the end of the chapter

1. Introduction

The combustion chamber is the most critical part of a gas turbine. The chamber had to be designed so that the combustion process to sustain itself in a continuous manner and the temperature of the products is sufficiently below the maximum working temperature in the turbine. In the conventional industrial gas turbine combustion systems, the combustion chamber can be divided into two areas: the primary zone and the secondary zone. The primary zone is where the majority of the fuel combustion takes place. The fuel must be mixed with the correct amount of air so that a stoichiometric mixture is present. In the secondary zone, unburned air is mixed with the combustion products to cool the mixture before it enters the turbine. In some design, there is an intermediate zone where help secondary zone to eliminate the dissociation products and burn-out soot.

The majority of the combustors are developed base on diffusion flames as they are very stable and fuel flexibility option. In a diffusion flame, there will be always stoichiometric regions regardless of overall stoichiometry. The main disadvantage of diffusion-type combustor is the emission as high temperature of the primary zone produced larger than 70 ppm NOx in burning natural gas and more than 100 ppm for liquid fuel [1]. Several techniques have been tried in order to reduce the amount of NOx produced in conventional combustors. In general, it is difficult to reduce NOx emissions while maintaining a high combustion efficiency as there is a tradeoff between NOx production and CO/UHC production.

In some recent installations, the premixed type of combustion has been selected to reduce NOx emissions bellow 10 ppm. Apart from the flame type change, there are some method such as "wet diffusion combustion", FGR[1] and SCR[2]. In an example of wet combustion, a nuzzle through which steam is injected is provided in the vicinity of the fuel injector. The level of NOx emission is controlled by the amount of steam. However, there is a limit on the increasing the steam flow rate as cause corresponding considerable CO emission. Furthermore, preparing pure steam in the required injection condition increases operational costs. Nowadays, wet combustion rarely applies due to water consumption and the penalty of reduced efficiency. Post Combustion treatments such as SCR are those which convert NOx compounds to nitrogen or absorb them from flue gas. These methods are relatively inexpensive to install but does not achieve NOx removal levels better than modern gas turbine combustor.

In this chapter, a short introduction of combustion process and then a description of some new pioneer combustor have been presented. As gas turbine manufacturers are looking for continuous operation or stable combustion, satisfactory emission level, minimum pressure loss and durability or life. Hence, the advanced combustor might include all of these criteria, so some of them are selected to discuss in details.

2. The combustion process

2.1. Type of combustion chamber

The diffusion and premixed flame are two main type of combustion, which are using in gas turbines. Apart from type of flame, there are two kind of combustor design, annular and tubular. The annular type mostly recommended in the propulsion of aircraft when small cross section and low weight are important parameters. Can or tubular combustors are cheaper and several of them can be adjusted for an industrial engine identically. Although there are different types of combustors, but generally, all combustion chambers have a diffuser, a casing, a liner, a fuel injector and a cooling arrangement. An entire common layout is visualized in figure 1.

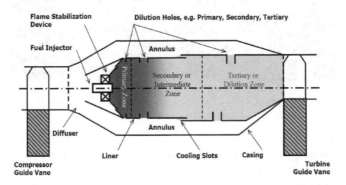

Figure 1. The layout of the combustion chamber.

---

1 Flue Gas Recirculation

2 Selective Catalytic Reduction

2.2. Flame stabilization

After the fuel has been injected into the air flow, the flow will enter the flame region. It does this with quite a high velocity, so to make sure the flame isn't blown away; suitable flame stabilization techniques must be applied. First, the high velocity of flow will be responsible for a pressure drop[3]. Secondly, the flame in the combustion chamber cannot survive if the air has a high velocity. So combustion chambers benefit from diffusers to slow down the air flow. There are two normal kinds of flame stabilizers: bluff-body flame holders and swirlers.

The shape of the bluff–body flame holder affects the flow stability characteristics through the influence on the size and shape of the wake region. Since the flame stabilization depends on size of the zone of recirculation behind the bluff–body, different geometries such as triangular, rectangular, circular and more complex shapes are being use. One of the basic problem of bluff-body flame holders is a considerable effect on pressure loss. Figure 2 shows a high speed image of three flame holders in atmospheric condition.

Figure 2. High speed images of the circular cylinder (top), square cylinder (middle) and V gutter (bottom) at Re = 30,000 and stoichiometric mixture [2].

Flow reversal can be applied in the primary zone. The best way to reverse the flow is to swirl it through using swirlers. The two most important types of swirlers are axial and radial. The advantage of flow reversal is that the flow speed varies a lot. So there will be a point at which the airflow velocity matches the flame speed where a flame could be stabilized. The degree of swirl in the flow is quantified by the dimensionless parameter, Sn known as the swirl number which is defined as:

---

3 This pressure drop is named the cold loss.

$$Sn = \frac{G_\theta}{G_x r} \tag{1}$$

Where:

$$G_\theta = \int_0^\infty (\rho u w + \overline{\rho u' w'}) r^2 dr$$

$$G_x = \int_0^\infty (\rho u^2 + \overline{\rho u'^2} + (p - p_\infty)) r dr$$

As this equation requires velocity and pressure profile of fluid, researchers proposed various expressions for calculating the swirl number. Indeed, the swirl number is a non-dimensional number representing the ratio of axial flux of angular momentum to the axial flux of axial momentum times the equivalent nozzle radius [3]. Tangential entry, guided vanes and direct rotation are three principal methods for generating swirl flow.

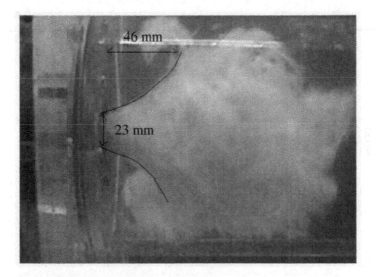

Figure 3. Photo of 60° flat guided vane swirler [4].

2.3. Type of flame

Most of the literatures divide combustible mixture into three categories as premixed, non-premixed and partially premixed combustion. If fuel and oxidizer are mixed prior ignition, then premixed flame will propagate into the unburned reactants. If fuel and air mix at the same time and same place as they react, the diffusion or non-premixed combustion will appear. Partially premixed combustion systems are premixed flames with non-uniform fuel-oxidizer mixtures.

Gas turbines' manufacturers traditionally tend to use diffusion flame where fuel mixes with air by turbulent diffusion and the flame front stabilized in the locus of the stoichiometric mixture. The temperature of reactant is as high as 2000 °C, so the acceptable temperature at the combustor walls and turbine blades would be provide by diluted air. Although the non-premixed mixture in gas turbine combustors shows more stability in operation than premixed mixtures, but their shortcoming is high level of nitrogen oxide emission. Two most common ways of emission reduction are water injection and catalytic converter. However, the former technique is not capable of reducing NOx to the expected level at many sites, while SCR adds complexity and expense to any project.

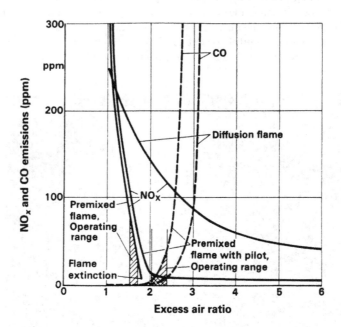

Figure 4. Operating range of premixed flames [5].

The idea of Dry Low NOx (DLN) systems proposed base on lean premixed combustion to reduce flame temperature by a non-stoichiometric mixture. Premixed systems can be operated at a much lower equivalence ratio such that the flame temperature and thermal NOx production throughout the system are decreased comparing with a diffusion system. The disadvantage of premixed systems is flame stability, especially at low equivalence ratios. Also, there is a tendency for the flame to flashback. Indeed, the current challenge of GT's developers is proposing a fuel flexible combustor for a stable combustion in all engine loads. The narrow range of fuel/air mixtures between the production of excessive NOx and excessive CO is illustrated in figure 4. NOx reduces by lowering flame temperature in a leaner mixture but CO, and unburned hydrocarbons (UHC) would increase contradictorily.

By increasing combustion residence time (volume) and preventing local quenching, CO and UHC will dissociate to CO2 and the other products. CO burns away more slowly than the other radicals, so to obtain very low level emission such as 10 ppm; it requires over 4 ms. As shown in figure 5, below 1100 °C the CO reaction becomes too slow to effectively remove the CO in an improved combustion chamber. The residence time usually does not change much on part-load because the normalized flow approximately remains constant with a variable loading.

$$NF = \frac{\dot{m}\sqrt{T}}{P} \tag{2}$$

Where $\dot{m}$ is the mass flow, T is combustion bulk temperature and P is combustor pressure. This will set a lower limit for the length of the primary zone in a DLN combustion system.

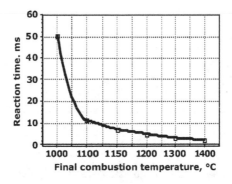

Figure 5. Calculated reaction time to achieve a CO concentration of 10 ppm in a commercial gas turbine exhaust [6].

2.4. Fuel

One of the features of heavy-duty gas turbines is a wide fuel capability. They can operate with vast series of commercial and process by-product fuels such as natural gas, petroleum distillates, gasified coal or biomass, gas condensates, alcohols, ash-forming fuels. In a review article, Molière offered essential aspects of fuel/machine interactions in thermodynamic performance, combustion and gaseous emission [7]. To sequester and store the CO2 of fossil fuel, some new research projects aim to assess the combustion performances of alternative fuels for clean and efficient energy production by gas turbines. Another objective is to ex-

tend the capability of dry low emission gas turbine technologies to low heat value fuels produced by gasification of biomass and H2 enriched fuels [8-10]. Significant quantity of
hydrogen in fuel has the benefit of high calorific value, but the disadvantage of high flame
speed and very fast chemical times. To classify gas turbine's fuels, a common way is to split
them between gas and liquid fuels, and within the gaseous fuels, to split by their calorific
value as shown in table 1.

	Typical composition	Lower Heating Value kJ/Nm³	Typical specific fuels
Ultra/Low LHV gaseous fuels	$H_2 < 10\%$ $CH_4 < 10\%$ $N_2 + CO > 40\%$	< 11,200 (< 300)	Blast furnace gas (BFG), Air blown IGCC, Biomass gasification
High hydrogen gaseous fuels	$H_2 > 50\%$ $C_xH_y = 0\text{-}40\%$	5,500-11,200 (150-300)	Refinery gas, Petrochemical gas, Hydrogen power
Medium LHV gaseous fuels	$CH_4 < 60\%$ $N_2 + CO_2 = 30\text{-}50\%$ $H_2 = 10\text{-}50\%$	11,200-30,000	Weak natural gas, Landfill gas, Coke oven gas, Corex gas
Natural gas	$CH_4 = 90\%$ $C_xH_y = 5\%$ Inert = 5%	30,000-45,000	Natural gas Liquefied natural gas
High LHV gaseous fuels	CH_4 and higher hydrocarbons $C_xH_y > 10\%$	45,000-190,000	Liquid petroleum gas (butane, propane) Refinery off-gas
Liquid fuels	C_xH_y, with x > 6	32,000-45,000	Diesel oil, Naphtha Crude oils, Residual oils, Bio-liquids

Table 1. Classification of fuels [11].

3. New combustion systems for gas turbines

Next-generation gas turbines will operate at higher pressure ratios and hotter turbine inlet
temperatures conditions that will tend to increase nitrogen oxide emissions. To conform to
future air quality requirements, lower-emitting combustion technology will be required. In
this section, a number of new combustion systems have been introduced where some of
them could be found in the market, and the others are under development.

3.1. Trapped vortex combustion (TVC)

The trapped vortex combustor (TVC) may be considered as a promising technology for both
pollutant emissions and pressure drop reduction. TVC is based on mixing hot combustion
products and reactants at a high rate by a cavity stabilization concept. The trapped vortex

combustion concept has been under investigation since the early 1990's. The earlier studies of TVC have been concentrated on liquid fuel applications for aircraft combustors [12].

The trapped vortex technology offers several advantages as gas turbines burner:

- It is possible to burn a variety of fuels with medium and low calorific value.

- It is possible to operate at high excess air premixed regime, given the ability to support high-speed injections, which avoids flashback.

- NOx emissions reach extremely low levels without dilution or post-combustion treatments.

- Produces the extension of the flammability limits and improves flame stability.

Flame stability is achieved through the use of recirculation zones to provide a continuous ignition source which facilitates the mixing of hot combustion products with the incoming fuel and air mixture [13]. Turbulence occurring in a TVC combustion chamber is "trapped" within a cavity where reactants are injected and efficiently mixed. Since part of the combustion occurs within the recirculation zone, a "typically" flameless regime can be achieved, while a trapped turbulent vortex may provide significant pressure drop reduction [14]. Besides this, TVC is having the capability of operating as a staged combustor if the fuel is injected into both the cavities and the main airflow. Generally, staged combustion systems are having the potential of achieving about 10 to 40% reduction in NOx emissions [15]. It can also be operated as a rich-burn, quick-quench lean-burn (RQL) combustor when all of the fuel is injected into the cavities [16].

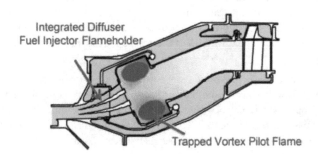

Figure 6. Trapped vortex combustor schematic.

An experiment in NASA with water injected TVC demonstrated a reduction in NOx by a factor three in a natural gas fueled and up to two in a liquid JP-8 fueled over a range in water/fuel and fuel/air ratios [17]. Replacement of natural gas fuel with syngas and hydrogen fuels has been studied numerically by Ghenai et al. [18]. The effects of secondary air jet momentum on cavity flow structure of TVC have been studied recently by Kumar and Mishra [19]. Although the actual stabilization mechanism facilitated by the TVC is relatively simple, a number of experiments and numerical simulations have been performed to enhance the stability of reacting flow inside trapped vortex. Xing et al. experimentally investigated lean blow-out of several combustors and the performance of slight temperature-raise in a single trapped vortex [20, 21]. In an experimental laboratory research, Bucher et al. proposed a new design for lean-premixed trapped vortex combustor [22].

3.2. Rich burn, quick- mix, lean burn (RQL)

Lean direct injection (LDI) and rich-burn/quick-quench/lean-burn (RQL) are two of the prominent low-emissions concepts for gas turbines. LDI operates the primary combustion region lean, hence, adequate flame stabilization has to be ensured; RQL is rich in the primary zone with a transition to lean combustion by rapid mixing with secondary air downstream. Hence, both concepts avoid stoichiometric combustion as much as possible, but flame stabilization and combustion in the main heat release region are entirely different. Relative to aviation engines, the need for reliability and safety has led to a focus on LDI of liquid fuels [23]. However, RQL combustor technology is of growing interest for stationary gas turbines due to the attributes of more effectively processing of fuels with complex composition. The concept of RQL was proposed in 1980 as a significant effort for reducing NOx emission [24].

It is known that the primary zone of a gas turbine combustor operates most effectively with rich mixture ratios so, a "rich-burn" condition in the primary zone enhances the stability of the combustion reaction by producing and sustaining a high concentration of energetic hydrogen and hydrocarbon radical species. Secondly, rich burn conditions minimize the production of nitrogen oxides due to the relative low temperatures and low population of oxygen containing intermediate species. Critical factors of a RQL that need to be considered are careful tailoring of rich and lean equivalence ratios and very fast cooling rates. So the combustion regime shifts rapidly from rich to lean without going through the high NOx route as shown in figure 7. The drawback of this technology is increased hardware and complexity of the system.

The mixing of the injected air takes the reaction to the lean-burn zone and rapidly reduces their temperature as well. On the other hand, the temperature must be high enough to burn CO and UHC. Thus, the equivalence ratio for the lean-burn zone must be carefully selected to satisfy all emissions requirements. Typically the equivalence ratio of fuel-rich primary zone is 1.2 to 1.6 and lean-burn combustion occurs between 0.5 and 0.7 [25].

Turbulent jet in a cross-flow is an important characteristic of RQL; so many researches have been conducted to improve it. The mixing limitation in a design of RQL/TVC combustion system addressed by Straub et al. [26]. Coaxial swirling air discussed experimentally by Cozzi and Coghe [27]. Furthermore, an experimental study of the effects of elevated pressure and temperature on jet mixing and emissions in an RQL reported by Jermakian et al. [28]. Fuel flexible combustion with RQL system is an interest of turbine manufacturer. GE reported results of a RQL test stand in their integrated gasification combined cycle (IGCC) power plants program [29, 30]. The test of Siemens-Westinghouse Multi-Annular Swirl Burner (MASB) was successfully performed at the University of

Tennessee Space Institute in Tullahoma [31]. Others, such as references [32-35] utilize CFD to investigate the performance of RQL combustor.

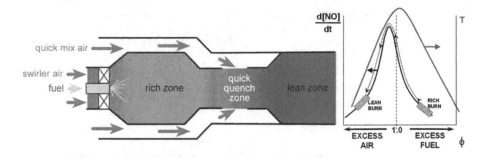

Figure 7. Rich-Burn, Quick-Mix, Lean-Burn combustor.

3.3. Staged air combustion

The COSTAIR[4] combustion concept uses continuously staged air and internal recirculation within the combustion chamber to obtain a stable combustion with low NOX and CO emissions. Research work on staged combustors started in the early 1970s under of the Energy Efficient Engine (E[3]) Program in the USA [36] and now widely used in industrial engines burning gaseous fuels, in both axial and radial configurations. The aero-derived GE LM6000 and CFM56-5B as well as RR211 DLE industrial engine employ staged combustion of premixed gaseous fuel/air mixtures. Recently, a research project proposed a COSTAIR burner system optimized for low calorific gases within a micro gas turbine [37].

The principle of staged air combustion is illustrated in Figure 8. It consists of a coaxial tube; the combustion air flows through the inner tube and the fuel through the outer cylinder ring. The combustion air is continually distributed throughout the combustion chamber by an air distributor with numerous openings on its contour, and fuel enters by several jets arranged around the air distributor.

The COSTAIR burner has the advantages of operating in full diffusion mode or in partially premixed mode. The heat is released more uniformly throughout the combustion chamber also the recirculated gas absorb some of the heat of combustion. It capable to work stable at cold combustor walls as well as high air ratio. Experimental measurements show that this combustion system allows clean exhaust. For instance, in an experimental research project of European Commission [39], NOx emission values was in the range of 2-4 ppm at an air ratio of 2.5 over different loading. Furthermore, the corresponding CO emission was less than 7 ppm.

---

4 COntinuous STaged Air

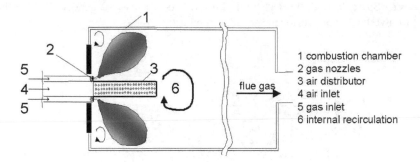

Figure 8. COSTAIR combustion concept [38].

Staged combustion can occur in either a radial or axial pattern, but in either case the goal is to design each stage to optimize particular performance aspects. The main advantages or major drawbacks of each type have been discussed by Lefebvre [25].

3.4. Mild combustion

Heat recirculating combustion was clearly described by Weinberg as a concept for improving the thermal efficiency [40]. In 1989, a surprising phenomenon was observed during experiments with a self-recuperative burner. At furnace temperatures of 1000°C and about 650°C air preheated temperature; no flame could be seen, but the fuel was completely burnt. Furthermore, the CO and NOx emissions from the furnace were considerably low [41]. Different combustion zones against rate of dilution and oxygen content is shown in figure 9. In flameless combustion, the oxidation of fuel occurs with a very limited oxygen supply at a very high temperature. Spontaneous ignition occurs and progresses with no visible or audible signs of the flames usually associated with burning. The chemical reaction zone is quite diffuse, and this leads to almost uniform heat release and a smooth temperature profile. All these factors could result in a much more efficient process as well as reducing emissions.

Flameless combustion is defined where the reactants exceed self-ignition temperature as well as entrain enough inert combustion products to reduce the final reaction temperature [42]. In the other word, the essence of this technology is that fuel is oxidized in an environment that contains a substantial amount of inert (flue) gases and some, typically not more than 3–5%, oxygen. Several different expressions are used to identify similar though such as HiTAC[5], HiCOT[6], MILD[7] combustion, FLOX[8] and CDC[9]. HiTAC refers to increase the air temperature by preheating systems such as regenerators. HiCOT commonly belongs to the

---

5 High Temperature Air Combustion

6 High-temperature Combustion Technology

7 Moderate or Intense Low-oxygen Dilution

8 FLameless OXidation

9 Colorless Distributed Combustion

wider sense, which exploits high-temperature reactants; therefore, it is not limited to air. A combustion process is named FLOX or MILD when the inlet temperature of the main reactant flow is higher than mixture autoignition temperature and the maximum allowable temperature increase during combustion is lower than mixture autoignition temperature, due to dilution [42]. The common key feature to achieve reactions in CDC mode (non-premixed conditions) is the separation and controlled mixing of higher momentum air jet and the lower momentum fuel jet, large amount of gas recirculation and higher turbulent mixing rates to achieve spontaneous ignition of the fuel to provide distributed combustion reactions [43]. Figure 10 schematically shows a comparison between conventional burner and flameless combustion.

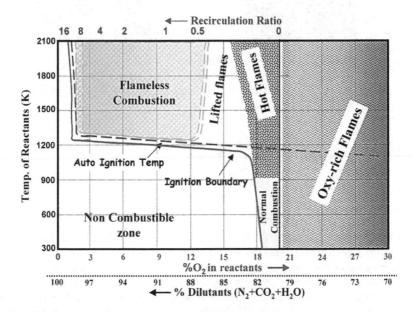

Figure 9. Different combustion regimes [11].

To recap, the main characteristics of flameless oxidation combustion are:

- Recirculation of combustion products at high temperature (normally > 1000 °C),
- Reduced oxygen concentration at the reactance,
- Low Damköhler number (Da^{10}),
- Low stable adiabatic flame temperature,
- Reduce temperature peaks,

---

10 A dimensionless number, equal to the ratio of the turbulence time scale to the time it takes chemical reaction.

- Highly transparent flame,
- Low acoustic oscillation and
- Low NOx and CO emissions.

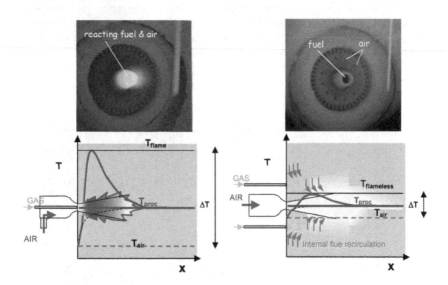

Figure 10. Flame (left) and flameless (right) firing.

In spite of a number of activities for industrial furnaces, the application of flameless combustion in the gas-turbine combustion system is in the preliminary phase [44]. The results from techno-economic analysis of Wang et al. showed that the COSTAIR and FLOX cases had technical and economic advantages over SCR [45]. Luckerath, R., et al., investigated flameless combustion in forward flow configuration in elevated pressure up to 20atm for application to gas turbine combustors [44, 46]. In a novel design of Levy et al. that named FLOXCOM, flameless concept has been proposed for gas turbines by establishing large recirculation zone in the combustion chamber [47, 48]. Lammel et al. developed a FLOX combustion at high power density and achieved low NOx and CO levels [49]. The concept of colorless distributed combustion has been demonstrated by Gupta et al. for gas turbine application in a number of publications [43, 50-55].

3.5. Surface stabilized combustion

One specification of gas turbine combustor is higher thermal intensity range (at least 5 MW/m³-atm) than industrial furnaces which operate at thermal intensity of less than 1 MW/m³-atm. Therefore, designs of gas turbine's combustors are based on turbulent flow concept, except a technology named NanoSTAR from Alzeta Corporation. Alzeta reported the proof-of-concept of high thermal intensity laminar surface stabilized flame by using a porous metal-fiber mat since 2001 [56-58]. Lean premixed combustion technology is limited by the apparition of combustion instabilities, which induce high pressure fluctuations, which can produce turbine damage, flame extinction, and CO emissions [59]. However, full scale test of NanoSTAR demonstrated low emissions performance, robust ignition and extended turndown ratio [60]. In particular, the following characteristics form the key specifications of NanoSTAR for distributed power generation gas turbine combustors [61]:

- The combustor fuel is limited to natural gas.

- Total combustor pressure drop limited to 2-4% of the system pressure.

- Operation at combustion air preheat temperatures up to 1150°F.

- Volumetric firing rates approaching 2 MMBtu/hr/atm/ft³.

- Turbine Rotor Inlet Temperatures (TRIT) over 2200°F (valid for the Mercury 50, although Allison has operated combustors at 2600°F).

- Operation with axial combustors or external can combustors.

- Expected component lifetimes of 30,000 hours for industrial turbines.

A single prototype burner Porous burner which sized to fit inside an annular combustion liner (about 2.5 inches in diameter by 7 inches in length) is shown in figure 11 with its arrangement in a typical combustor.

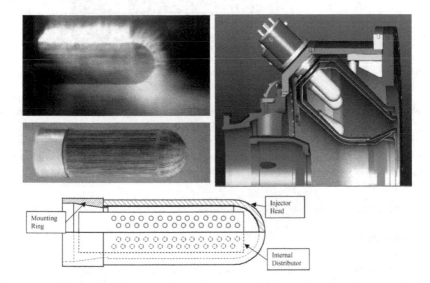

Figure 11. NanoSTAR burner and its arrangement in a canted combustion system [62].

The operation of this type of surface stabilized combustion is characterized by the schematic in Figure 12(left), which shows premixed fuel and air passing through the metal fiber mat in two distinct zones. Premixed fuel comes through the low conductivity porous and burns in narrow zones, A, as it leaves the surface. Under lean conditions this will manifest as very short laminar flamelets, but under rich conditions the surface combustion will become a dif-fusion dominated reaction stabilized just over a millimeter above the metal matrix, which proceeds without visible flame and heats the outer surface of the mat to incandescence. Sec-ondly, adjacent to these radiant zones, the porous plate is perforated to allow a high flow of the premixed fuel and air. This flow forms a high intensity flame, B, stabilized by the radiant zones so, it is possible to achieve very high fluxes of energy, up to 2MMBtu/hr/ft^2 [63]. A picture of an atmospheric burner in operation clearly shows the technology in action (right of figure 12).

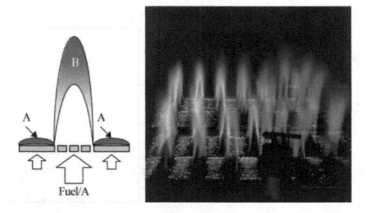

Figure 12. Surface stabilized burner pad firing at atmospheric conditions.

The specific perforation arrangement and pattern control the size and shape of the laminar flamelets. The perforated zones operate at flow velocities of up to 10 times the laminar flame speed producing a factor of ten stretch of the flame surface and resulting in a large laminar flamelets. The alternating arrangement of laminar blue flames and surface combustion, al-lows high firing rates to be achieved before flame liftoff occurs, with the surface combustion stabilizing the long laminar flames by providing a pool of hot combustion radicals at the flame edges.

4. Conclusion

A review of technologies for reducing NOx emissions as well as increasing thermal efficien-cy and improving combustion stability has been reported here. Trade-offs when installing low NOx burners in gas turbines include the potential for decreased flame stability, reduced operating range and more strict fuel quality specifications. In the other word, although, the turbine inlet temperature is the major factor determining the overall efficiency of the gas turbine but higher inlet temperatures will result in larger NOx emissions. So the essential requirement of new combustor design is a trade-off between low NOx and improved efficiency.

Author details

M. Khosravy el_Hossaini

Research Institute of Petroleum Industry, Iran

References

[1] Chambers A and Trottier S (2007) Technologies for Reducing Nox Emissions from Gas-Fired Stationary Combustion Sources. Alberta Research Council, Edmonton, Canada

[2] Kiel B, Garwick LK, Gord JR, Miller J, Lynch A, Hill R and Phillips S (2007) A Detailed Investigation of Bluff Body Stabilized Flames. 45th AIAA Aerospace Sciences Meeting and Exhibit.

[3] Gupta AK, Lilley DG and Syred N (1984) Swirl Flows, Abacus Press.

[4] Jaafar MNM, Jusoff K, Osman MS and Ishak MSA (2011) Combustor Aerodynamic Using Radial Swirler. International Journal of the Physical Sciences. 6: 3091 - 3098.

[5] Moore MJ (1997) Nox Emission Control in Gas Turbines for Combined Cycle Gas Turbine Plant. Proceedings of the Institution of Mechanical Engineers, Part A: Journal of Power and Energy. 211: 43-52.

[6] Kajita S and Dalla Betta R (2003) Achieving Ultra Low Emissions in a Commercial 1.4 Mw Gas Turbine Utilizing Catalytic Combustion. Catalysis Today. 83: 279-288.

[7] Molière M (2000) Stationary Gas Turbines and Primary Energies: A Review of Fuel Influence on Energy and Combustion Performances. International Journal of Thermal Sciences. 39: 141-172.

[8] Gupta KK, Rehman A and Sarviya RM (2010) Bio-Fuels for the Gas Turbine: A Review. Renewable and Sustainable Energy Reviews. 14: 2946-2955.

[9] Gökalp I and Lebas E (2004) Alternative Fuels for Industrial Gas Turbines (Aftur). Applied Thermal Engineering. 24: 1655-1663.

[10] Juste GL (2006) Hydrogen Injection as Additional Fuel in Gas Turbine Combustor. Evaluation of Effects. International Journal of Hydrogen Energy. 31: 2112-2121.

[11] Jones R, Goldmeer J and Monetti B (2011) Addressing Gas Turbine Fuel Flexibility. GE Energy

[12] Haynes J, Janssen J, Russell C and Huffman M (2006) Advanced Combustion Systems for Next Generation Gas Turbines. United States. Dept. of Energy, Washington, D.C.; Oak Ridge, Tenn.

[13] Sturgess GJ and Hsu KY (1998) Combustion Characteristics of a Trapped Vortex Combustor. Applied vehicle technology panel symposium.

[14] Bruno C and Losurdo M (2007) The Trapped Vortex Combustor: An Advanced Combustion Technology for Aerospace and Gas Turbine Applications. In: Syred N and Khalatov A, Syred N and Khalatov A, editors. Advanced Combustion and Aerothermal Technologies. Springer Netherlands, pp 365-384.

[15] Mishra DP (2008) Fundamentals of Combustion, Prentice-Hall Of India Pvt. Limited.

[16] Acharya S, Mancilla PC and Chakka P (2001) Performance of a Trapped Vortex Spray Combustor. ASME International Gas Turbine Conference. ASME.

[17] Hendricks RC, Shouse DT and Roquemore WM (2005) Water Injected Turbomachinery. NASA, Glenn Research Center

[18] Ghenai C, Zbeeb K and Janajreh I (2012) Combustion of Alternative Fuels in Vortex Trapped Combustor. Energy Conversion and Management. In press

[19] Ezhil Kumar PK and Mishra DP (2011) Numerical Simulation of Cavity Flow Structure in an Axisymmetric Trapped Vortex Combustor. Aerospace Science and Technology. In Press

[20] Xing F, Wang P, Zhang S, Zou J, Zheng Y, Zhang R and Fan W (2012) Experiment and Simulation Study on Lean Blow-out of Trapped Vortex Combustor with Various Aspect Ratios. Aerospace Science and Technology. 18: 48-55.

[21] Xing F, Zhang S, Wang P and Fan W (2010) Experimental Investigation of a Single Trapped-Vortex Combustor with a Slight Temperature Raise. Aerospace Science and Technology. 14: 520-525.

[22] Bucher J, Edmonds RG, Steele RC, Kendrick DW, Chenevert BC and Malte PC (2003) The Development of a Lean-Premixed Trapped Vortex Combustor. ASME Turbo Expo 2003 Power for Land, Sea, and Air.

[23] Dunn-Rankin D (2008) Lean Combustion: Technology and Control, Academic Press, USA.

[24] Mosier SA and Pierce RM (1980) Advanced Combustion Systems for Stationary Gas Turbine Engines. Volume I. Review and Preliminary Evaluation. Final Report December 1975-September 1976. pp Medium: X; Size: Pages: 49.

[25] Lefebvre AH and Ballal DR (2010) Gas Turbine Combustion: Alternative Fuels and Emissions, Taylor & Francis.

[26] Straub DL, Casleton KH, Lewis RE, Sidwell TG, Maloney DJ and Richards GA (2005) Assessment of Rich-Burn, Quick-Mix, Lean-Burn Trapped Vortex Combustor for Stationary Gas Turbines. Journal of engineering for gas turbines and power. 127: 36-41.

[27] Cozzi F and Coghe A (2012) Effect of Air Staging on a Coaxial Swirled Natural Gas Flame. Experimental Thermal and Fluid Science. In press

[28] Jermakian V, McDonell VG and Samuelsen GS (2012) Experimental Study of the Effects of Elevated Pressure and Temperature on Jet Mixing and Emissions in an Rql Combustor for Stable, Efficient and Low Emissions Gas Turbine Applications. Advanced Power and Energy Program, University of California, Irvine

[29] Feitelberg AS, Jackson MR, Lacey MA, Manning KS and Ritter AM (1996) Design and Performance of a Low Btu Fuel Rich-Quench-Lean Gas Turbine Combustor. Advanced coal-fired power systems review meeting. USA DOE, Morgantown Energy Technology Center.

[30] Feitelberg AS and Lacey MA (1998) The Ge Rich-Quench-Lean Gas Turbine Combustor. Journal of Engineering for Gas Turbines and Power, Transactions of the ASME. 120: 502-508.

[31] Brushwood J (1999) Syngas Combustor for Fluidized Bed Applications 15th Annual Fluidized Bed Conference.

[32] Howe GW, Li Z, Shih TI-P and Nguyen HL (1991) Simulation of Mixing in the Quick Quench Region of a Richburn-Quick Quench Mix-Lean Burn Combustor. 29th Aerospace Sci Meeting. AIAA.

[33] Cline MC, Micklow GJ, Yang SL and Nguyen HL (1992) Numerical Analysis of the Flow Fields in a Rql Gas Turbine Combustor.

[34] Talpallikar MV, Smith CE, Lai MC and Holdeman JD (1992) Cfd Analysis of Jet Mixing in Low Nox Flametube Combustors. Journal of Engineering for Gas Turbines and Power. 114: 416-424.

[35] Blomeyer M, Krautkremer B, Hennecke DK and Doerr T (1999) Mixing Zone Optimization of a Rich-Burn/Quick-Mix/Lean-Burn Combustor. Journal of Propulsion and Power 15: 288-303.

[36] Wulff A and Hourmouziadis J (1997) Technology Review of Aeroengine Pollutant Emissions. Aerospace Science and Technology. 1: 557-572.

[37] Leicher J, Giese A, Görner K, Scherer V and Schulzke T (2011) Developing a Burner System for Low Calorific Gases in Micro Gas Turbines: An Application for Small Scale Decentralized Heat and Power Generation International Gas Union Research Conference.

[38] Al-Halbouni A, Flamme M, Giese A, Scherer V, Michalski B and Wünning JG (2004) New Burner Systems with High Fuel Flexibility for Gas Turbines. 2nd International Conference on Industrial Gas Turbine Technologies.

[39] Flamme M (2004) New Combustion Systems for Gas Turbines (Ngt). Applied Thermal Engineering. 24: 1551-1559.

[40] WEINBERG F (1996) Heat-Recirculating Burners : Principles and Some Recent Developments. Combustion Science and Technology. 121: 3-22.

[41] Wünning J (2005) Flameless Oxidation. 6th HiTACG Symposium.

[42] Cavaliere A and de Joannon M (2004) Mild Combustion. Progress in Energy and Combustion Science. 30: 329-366.

[43] Arghode VK, Gupta AK and Bryden KM (2012) High Intensity Colorless Distributed Combustion for Ultra Low Emissions and Enhanced Performance. Applied Energy. 92: 822-830.

[44] Li P, Mi J, Dally B, Wang F, Wang L, Liu Z, Chen S and Zheng C (2011) Progress and Recent Trend in Mild Combustion. SCIENCE CHINA Technological Sciences. 54: 255-269.

[45] Wang YD, Huang Y, McIlveen-Wright D, McMullan J, Hewitt N, Eames P and Rezvani S (2006) A Techno-Economic Analysis of the Application of Continuous Staged-Combustion and Flameless Oxidation to the Combustor Design in Gas Turbines. Fuel Processing Technology. 87: 727-736.

[46] Luckerath R, Meier W and Aigner M (2008) Flox Combustion at High Pressure with Different Fuel Compositions. Journal of Engineering for Gas Turbines and Power. 130: 011505.

[47] Costa M, Melo M, Sousa J and Levy Y (2009) Experimental Investigation of a Novel Combustor Model for Gas Turbines. Journal of Propulsion and Power 25: 609-617.

[48] Levy Y, Sherbaum V and Arfi P (2004) Basic Thermodynamics of Floxcom, the Low-Nox Gas Turbines Adiabatic Combustor. Applied Thermal Engineering. 24: 1593-1605.

[49] Lammel O, Schutz H, Schmitz G, Luckerath R, Stohr M, Noll B, Aigner M, Hase M and Krebs W (2010) Flox Combustion at High Power Density and High Flame Temperatures. Journal of Engineering for Gas Turbines and Power. 132: 121503.

[50] Arghode VK and Gupta AK (2011) Development of High Intensity Cdc Combustor for Gas Turbine Engines. Applied Energy. 88: 963-973.

[51] Khalil AEE and Gupta AK (2011) Distributed Swirl Combustion for Gas Turbine Application. Applied Energy. 88: 4898-4907.

[52] Arghode VK and Gupta AK (2010) Effect of Flow Field for Colorless Distributed Combustion (Cdc) for Gas Turbine Combustion. Applied Energy. 87: 1631-1640.

[53] Arghode VK, Khalil AEE and Gupta AK (2012) Fuel Dilution and Liquid Fuel Operational Effects on Ultra-High Thermal Intensity Distributed Combustor. Applied Energy. 95: 132-138.

[54] Khalil AEE, Arghode VK, Gupta AK and Lee SC (2012) Low Calorific Value Fuelled Distributed Combustion with Swirl for Gas Turbine Applications. Applied Energy. 98: 69-78.

[55] Khalil AEE and Gupta AK (2011) Swirling Distributed Combustion for Clean Energy Conversion in Gas Turbine Applications. Applied Energy. 88: 3685-3693.

[56] Greenberg SJ, McDougald NK and Arellano LO (2004) Full-Scale Demonstration of Surface-Stabilized Fuel Injectors for Sub-Three Ppm Nox Emissions. ASME Conference Proceedings. 2004: 393-401.

[57] Greenberg SJ, McDougald NK, Weakley CK, Kendall RM and Arellano LO (2003) Surface-Stabilized Fuel Injectors with Sub-Three Ppm Nox Emissions for a 5.5 Mw Gas Turbine Engine. International Gas Turbine and Aeroengine Congress and Exhibition. American Society of Mechanical Engineers.

[58] Weakley CK, Greenberg SJ, Kendall RM, McDougald NK and Arellano LO (2002) Development of Surface-Stabilized Fuel Injectors with Sub-Three Ppm Nox Emissions. International Joint Power Generation Conference. American Society of Mechanical Engineers.

[59] Cabot G, Vauchelles D, Taupin B and Boukhalfa A (2004) Experimental Study of Lean Premixed Turbulent Combustion in a Scale Gas Turbine Chamber. Experimental Thermal and Fluid Science. 28: 683-690.

[60] Mcdougald NK (2005) Development and Demonstration of an Ultra Low Nox Combustor for Gas Turbines. USA DOE, Office of Energy Efficiency and Renewable Energy, Washington, D.C; Oak Ridge, Tenn.

[61] Arellano LO, Bhattacharya AK, Smith KO, Greenberg SJ and McDougald NK (2006) Development and Demonstration of Engine-Ready Surface-Stabilized Combustion System. ASME Turbo Expo 2006: Power for Land, Sea, and Air.

[62] Arellano L, Smith KO, California Energy Commission. Public Interest Energy R and Solar Turbines I (2008) Catalytic Combustor-Fired Industrial Gas Turbine Pier Final Project Report, California Energy Commission, [Sacramento, Calif.].

[63] Clark H, Sullivan JD, California Energy Commission. Public Interest Energy R, California Energy Commission. Energy Innovations Small Grant P and Alzeta C (2001) Improved Operational Turndown of an Ultra-Low Emission Gas Turbine Combustor, California Energy Commission, Sacramento, Calif.

Fault Detection in Systems and Materials

Chapter 8

System Safety of Gas Turbines: Hierarchical Fuzzy Markov Modelling

G. G. Kulikov, V. Yu. Arkov and A.I. Abdulnagimov

Additional information is available at the end of the chapter

1. Introduction

Reliability, safety and durability represent important properties of modern aircraft, which is necessary for its effective in-service use.

The reason of the main hazard for aircraft are both random and determined negative influences rendering the controlled object during its use. Faults, failures, disturbances, noises, influences of environment and control errors represent the objectively existing stream of random negative influences on the object.

Statistically, in the recent years the majority of aircraft incidents are connected with the human factor and late fault detection in plane systems. In this regard, requirements to flight safety which demand development of new methods and algorithms of control-and-condition monitoring/ diagnostic for complex objects raise every year. The analysis of modern gas turbine engines has shown that most faults appears in the engine itself and its FADEC (40-75% for FADEC, Figure 1).

The percentage of faults for FADEC depends on the achieved values for no-failure operation indicators of the engine and FADEC.

During the development of FADEC, it is necessary to adhere to the principles and methods guaranteeing safety and reliability of aircraft in use to guarantee proper responses in all range of negative influences.

Full information on its work is necessary for complete control of a condition of the engine:

1. Reliable detection of a fault cause providing decision-making on a technical condition of gas turbines;

2.   Reliable diagnosis and localization of faults and negative influences are necessary for definition of technical condition of gas turbines for the purpose of providing a reconfiguration and functioning of its subsystems [1, 2].

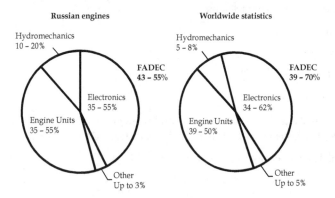

Figure 1. Faults percentage for engines and FADEC

The hardware for condition monitoring of measurement channels in many cases allows to detect only catastrophic (breakage or short circuit) faults, i.e. their stochastic properties on time of the process observed in one object and on a set of objects are not distinguishable [3]. The criteria of warning messages on faults appearance are based mainly on determined logic operations and distinguish between only two conditions: "operational" (fully operational) or "fault". In this chapter, hierarchical fuzzy Markov models for quantitative estimation of system safety of gas turbines taking into account the monitoring of cause-effect relations are considered. Transition from two-valued to fuzzy logic for estimation of degradation indexes and the analysis of fault developments for the gas turbine and its FADEC is considered for this purpose.

2. Hierarchical model of faults development processes in gas turbines

Complex diagnostics of the power plant is proposed to be carried out on elements and units, using the hierarchy analysis method [4, 5]. First, decomposition into independent subsystems of various hierarchy levels is carried out on structural features. Similarly, the power plant and its systems are represented in the form of hierarchy of elements and blocks.

This approach enables cause-effect relationships to be identified on the hierarchy structure of a system.

In Figure 2, the hierarchical structure of states of the power-plant is shown. The power plant is represented in the form of a hierarchical structure as the complex system consisting of subsystems and elements (units) with built-in test/monitoring functions, according to the distributed architecture. For this purpose, the power plant decomposition might be performed into independent subsystems with various levels of hierarchy on structural and functional features in the following way:

- Control and monitoring system (FADEC);
- Hydro mechanical system (actuators);
- Fuel system;
- Start-up system;
- Lubricant oil system;
- Drainage system, etc.

The hierarchy analysis allows to utilize the state model on the basis of faults development which enables the system state to be estimated at each level of the hierarchy.

The mathematical model of states is represented as

$$S = <G, F, L, R>,$$

where S is state vector,

   G is hierarchy of system faults,

   F is quantitative estimate of faults,

   L is set of fault influence indexes,

   R is mutual influence system of faults.

The depth of hierarchy G is referred to as h, and $h = 0$ for the root element of G.

For G the following conditions are satisfied:

1. There is splitting of G into subsets of h_k, $k = 1 \dots n$.

2. From $x \in L_k$ follows that $x^- \subset h_{k+1}$, $k = 1, \dots, n-1$.

3. From $x \in L_k$ follows that $x^+ \subset h_{k-1}$, $k = 2, \dots, n$.

For every $x \in G$ there is a weight function such as:

$$\omega_x : x^- \to [0, 1]; \text{ where } \sum_{y \in x^-} \omega_x(y) = 1.$$

The sets of h_i are the hierarchy levels, and function ω_x is a function of fault priority of one level concerning the state of the power-plant x. Notice that if $x^- \not\subset h_{k+1}$ (for some level of h_w), then ω_x can be defined for all h_k, if it equals to zero for all faults in h_{k+1} which do not belong to x^-.

The hierarchical FADEC model integrates:

- functional structure (block diagram);
- physical structure;

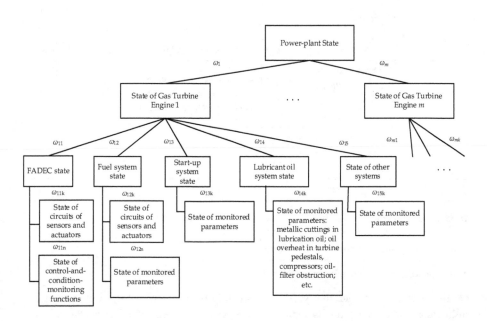

Figure 2. Hierarchical state structure of power plant

Fault levels	Faults		State	Fault handling priority
Level 1	FADEC fault		Fault	Immediate
	Fault of lane B			
Levels 2-4	Fault of control function (B)			
	Fault of control loop (B)			
Levels 5-6	Integrated fault of measurement for alternative control law	Fault of actuator control circuit (B)		
Level 7	Fault of lane A		Degradation	Short-term
Level 8	Fault of control function (A)			
	Fault of control loop	Fault of control loop (A)		
Level 9	Integrated fault of measurement (fault of same measured parameters in channels A)	Fault of actuator control circuit (A) (fault of communication lines of sensors or fault of FADEC hardware)		
Level 10	Fault of measurement in channel (break or short circuit)			Long-term
	Sensors	Actuators		

Table 1. Fragment of hierarchical classification of faults for FADEC

- tree states (state structure) of elements and units;

- tree of failures influence indexes.

On the hierarchical model, the system of faults interference R with logical operations of a disjunction and a conjunction is applied. Such a system of faults interference allows to analyze the state of all power-plant, both from the bottom up to the top, and from the top down to the bottom and to carry out deeper analysis on various levels of decomposition of the control system using an intermediate state: degradation.

The degradation is understood as "package/complex of degradationary changes of the system" and the degradationary change is "a separately considered irreversible change of a structure of the system, worsening its properties, changing the parameters and characteristics".

Define the main faults of FADEC, the priorities of their elimination and the state they belong to. In Table 1, an exemplary of a fragment of the hierarchical classification of faults for sensors and actuators of FADEC is presented.

Fault levels 1 through 6 demand immediate handlings and correspond to the "catastrophic" and "critical" states by FAA classification (Federal Aviation Administration, U.S. Department of Transportation), given in [6]. The emerging of such states requires immediate landing of the aircraft. Fault levels 7 through 9 are classified as a "marginal" state and demand operative handling after landing. In this case, it is possible to continue the flight, but post-flight repair on the ground is required. Faults at the 10th level demand their handling in long-term prospect.

The degradation process for FADEC starts at the 10th level of hierarchy. From the 4th level of hierarchy, the system starts to approach the system crash that can be regarded as «a critical situation».

Note that development of such faults in certain cases can be detected in advance by estimating the states of elements not only at the level of "0-1" (fully operational, operational/working, fault), but also by considering their gradual degradation.

The state of an element or a system is proposed to be represented in the form of three parameters { operational, degradation, fault }, see Tab. 2.

In the operational state $S = 0$, while during fault $S = 1$. The degradation degree range from "0" to "1". Thus, the extreme values "0" and "1" are defined according to the determined logic, which is realized in the conventional FADEC (according to the design specifications for the system). The introduction of this intermediate state of "degradation" expands the informativity of the conventional condition monitoring algorithms.

Operational	Degradation	Fault
$S = 0$	$0 < S < 1$	$S = 1$

Table 2. Fuzzy representation of state

Based on the faults analysis and the hierarchy of states of the system at each level, the degree of degradation of each item or sub-unit is determined (Figure 3). Fault states are classified via degradation degree as "Negligible", "Marginal", "Critical" and "Catastrophic" [6]. The estimation of the degradation degree is defined on the membership function S which takes values in the range of $S \in [0, 1]$. If the degradation degree is closer to "1", the distance to a critical situation will be closer. If the analysis of a system showed that the state vector is { 0,1 0,6 0,3 }, it is possible to ensure that there is a "distance" before complete fault (a critical situation). As soon as the system state will worsen with the appearance of new faults and will give the following state vector { 0 0,3 0,7 }, then there will be a distance of 0,3 to a system crash. Thus the most informative indicator will be a tendency of faults appearance (trend), not the existence of degradation itself. Visual trend analysis provides an estimate of time before the critical situations develop and, thus, for early planning of the crew actions [7, 8].

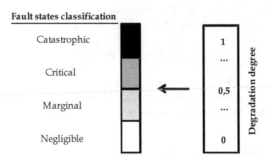

Figure 3. Estimation of "degradation" state

Consider an example of correspondence of degradation degree and the operational state. At the degradation degree of 0,25, the system is capable to carry out 75% of demanded functions (50% at 0,5 degradation, 25% at 0,75 and 0% at 1, which is the unavailable state). Such scale allows to define a "threshold" state, below which further operation is not allowed for safety reasons. Using the degradation degree, it is also possible to estimate the distance to a critical situation and the speed of approximation to it (Figure 4).

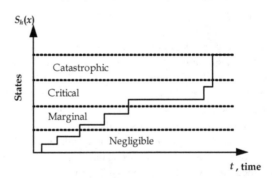

Figure 4. Trend of state dynamics during flight

Thus, the hierarchical model of fault developments allows to decompose the power-plant on hierarchy levels for obtaining quantitative estimates of the degradation state and gradual faults. The hierarchy analysis allows to utilize the state model and to estimate the system state at each level of hierarchy. The state is represented in the form of a vector with parameters { operational, degradation, fault }. Depending on the degradation degree it is possible to make an estimation of operability of object and system safety.

3. Fuzzy technique of determination of state parameters of gas turbine and its systems

The built-in monitoring system (BMS) is a subsystem of monitoring, diagnosis and classification of faults of the gas turbine and its systems. The fault existence corresponds to a logic state of "1", the absence does to a logic state of "0". Such state classification doesn't allow to establish a "prefault" state, to trace faults' development, and to define degradation of the system and its elements. For more detailed analysis, the estimation of the intermediate state of degradation is proposed. For this purpose, the use of fuzzy logic is considered. Signals from sensors, and also logic state parameters from BMS will transform to linguistic variables during fuzzification to a determined value arrives to the input of the fuzzifier. Let x is the state parameter of an element (for example, the sensor). It is necessary to define fuzzy spaces of input and output variables, and also terms for FADEC sensors. All signals from sensors and actuators will transform into linguistic variables by fuzzification.

Consider an example of fuzzy representation of a state of the two-channel sensor of rotational speed of a high pressure turbocompressor rotor. For monitoring of operational condition of communication lines of the FADEC sensor, a linguistic variable is introduced in the following way:

$$\Omega = <x_n, \; B=(x_n), \; U, \; G, \; M>,$$

where Ω_n is the sensor state,

x_n is the number of events when n is beyond the allowed limit band;

B is { operational, fault };

U is $[0,4]$

G is the syntactic rule generating terms of set B,

M is the semantic rule, which to each linguistic value x associate with its sense of $M(x_n)$, and $M(x_n)$ designates a fuzzy subset of the carrier U.

Say that the sensor is considered failed after the fourth appearance of the shaft speed measurement beyond the allowed boundary, therefore the membership function is formed as shown in Figure 5.

At a single appearance out of limit ($x_n = 1$), membership function B_1 takes the value 0,7, and $B_2 = 0,3$. The degradation degree takes the value of membership function B_2. If the repeated breaking the limit doesn't prove to be true during the set period of time, the monitoring algorithm cancels the measurement: $B_1 = 1$, $B_2 = 0$.

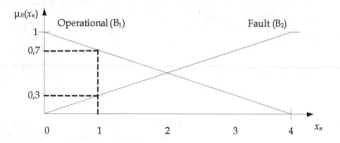

Figure 5. Membership functions of "sensor state"

In Figure 5 two membership functions are shown: state B1 corresponds to the function μ_{B1} (x_n), B_2 is described by the function μ_{B2} (x_n).

The way of creating fuzzy rules is presented in Figure 6. This rule base is represented by the table, which is filled in with fuzzy rules as follows [9]:

$$R^{(1)} : IF(x_n = A_1 \ AND \ x_n = B_1) \ THEN \ y = T_1$$

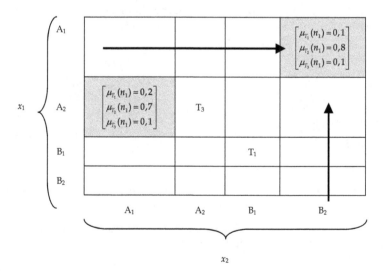

Figure 6. Example of fuzzy rule base

The values μ_{T1}, μ_{T2}, μ_{T3} are set in the cell at the row "Operational" (A_1) and the column "Fault" (B_2).

Consider a fragment of the rule base for estimates of the sensor state of the low pressure shaft speed. Formulate the first rule: if the 1st coil (n_{11}) of the sensor is operational (A_1) and the 2nd coil (n_{12}) of the sensor is operational (A_1), then the sensor is in the operational condition with the following membership functions:

$$[\mu_{\text{operational}}(n_1)=1; \quad \mu_{\text{degradation}}(n_1)=0; \quad \mu_{\text{fault}}(n_1)=0].$$

Write down this rule as follows:

$$R^{(1)}: IF\,(n_{11} = A_1\,AND\,n_{12} = A_1)\,THEN\,\,y = T_1 \Rightarrow$$
$$[\mu_{T_1}(n_1)=1; \quad \mu_{T_2}(n_1)=0; \quad \mu_{T_3}(n_1)=0].$$

Other rules are created in the similar way.

Given a greater number of possible conditions (for example, greater number of the duplicated coils of the sensor), one can develop a discrete-ordered scale of state parameters (Figure 7).

For further analysis of the system, enter the faults influence indexes at each level of hierarchy, using a method of pairwise comparison as it is carried out in the hierarchy analysis method.

Quantitative judgements on the importance of faults are performed for each pair of faults (F_i, F_j) and these are represented by matrix A of the $n \times n$ size.

$$A = (a_{ij}), \quad (i,\ j = 1, 2, 3).$$

where a_{ij} is the relative importance of fault F_i in regard to F_j. The value A_{ij} defines the importance (respective values) F_i of faults in comparison with F_j.

Elements a_{ij} are defined by the following rules:

1. If $a_{ij} = \alpha$, $a_{ji} = 1/\alpha$, $\alpha \ne 0$.

2. If fault F_i has identical relative importance with F_j, then $a_{ij} = 1$, $a_{ji} = 1$, in particular $a_{ii} = 1$ for all i.

Thus, a back-symmetric matrix A is obtained:

$$A = \begin{bmatrix} 1 & a_{12} & \cdots & a_{1n} \\ 1/a_{12} & 1 & \cdots & a_{2n} \\ \cdots & \cdots & \cdots & \cdots \\ 1/a_{1n} & 1/a_{2n} & \cdots & 1 \end{bmatrix}. \qquad (\text{ID2})$$

After the representation of quantitative judgements about the fault pairs (F_i, F_j) in a numerical expression with the numbers a_{ij}, the problem is reduced to that n possible faults F_1, F_2, ..., F_n will receive a corresponding set of numerical weights ω_1, ω_2, ..., ω_n, which would reflect the fixed judgements about the condition of the gas turbine subsystem.

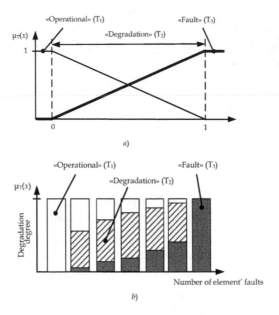

Figure 7. Continuous (a) and discrete (b) scale of degradation

If the expert judgement is absolute at all comparisons for all i, j, k, then matrix A is called *consistent*.

If the diagonal of matrix A consists of units ($a_{ij} = 1$) and A is the consistent matrix, then at small changes in a_{ij} the greatest eigenvalue λ_{max} is close to n, and the other eigenvalue are close to zero.

Based on the matrix of pair comparison values of faults A, the vector of priorities for fault classification is obtained, along with vector ω satisfying the criterion:

$$A\omega = \lambda_{max}\omega,$$

where ω is the eigenvector of matrix A and λ_{max} is the maximum eigenvalue, which is close to the matrix order n.

As it is desirable to have the normalized solution, let's slightly change ω, considering $\alpha = \sum\limits_{i=1}^{n} \omega_i$ and replacing ω with $(1/\alpha)\,\omega$. This provides uniqueness, and also that $\sum\limits_{i=1}^{n} \omega_i = 1$.

$$
A = \begin{array}{c} \\ F_1 \\ F_2 \\ \vdots \\ F_n \end{array}
\begin{array}{cccc}
F_1 & F_2 & \cdots & F_n
\end{array}
\left[\begin{array}{cccc}
\omega_1/\omega_1 & \omega_1/\omega_2 & \cdots & \omega_1/\omega_n \\
\omega_2/\omega_1 & \omega_2/\omega_2 & \cdots & \omega_2/\omega_n \\
\vdots & \vdots & \vdots & \vdots \\
\omega_n/\omega_1 & \omega_n/\omega_2 & \cdots & \omega_n/\omega_n
\end{array}\right]
\left[\begin{array}{c}
\omega_1 \\ \omega_2 \\ \vdots \\ \omega_n
\end{array}\right]
= \lambda_{max}
\left[\begin{array}{c}
\omega_1 \\ \omega_2 \\ \vdots \\ \omega_n
\end{array}\right].
$$

Note that small changes in a_{ij} cause small change in λ_{max}, then the deviation of the latter from n is a coordination measure. It allows estimating proximity of the obtained scale to the basic scale of relations. Hence, the coordination index

$$
(\lambda_{max}-n)/(n-1)
$$

is considered to be an indicator of "proximity to coordination". Generally, if this number is not greater than 0.1 then it is possible to be satisfied with the judgements about the faults importance.

At each level h_i of the hierarchy for n elements of the gas turbine and its subsystems, the state vector {operational, degradation, fault} is determined, taking into account the influence coefficients of failures:

$$
S_{h_i}(x_n) = \mu_{h_i}(x_n) \cdot \omega_i,
$$

where $\mu_{h_i}(x_n)$ is the membership function value of the element x_n (degradation degree). To determine the element/unit state of the hierarchy at a higher level $S_{h_i}(x_n)$ for the input states of low-level $S_{h_{i-1}}(x_n)$ one stage of defuzzification is performed.

The output value $S_{h_i}(x_n)$ is presented in the form of the determined vector of state with parameters { operational, degradation, fault }.

The state estimation begins with the bottom level of hierarchy. The description of a state set obtained by means of fuzzification and deffuzification with the use of the logic operations of disjunction ∨ (summing), and conjunction ∧ (multiplication), which are designated as follows:

———————————   logic operation "AND"

▬ ▬ ▬ ▪   logic operation "OR"

In performing operation "AND" in the inference system of fuzzy logic, the terminal tops are summed in order to determine the general state at one level of hierarchy that is presented as follows:

$$
X_\Sigma = x_1 \vee x_2 \vee \ldots \vee x_n
$$

In performing operation "OR", the "worst" state vector is chosen, with the maximum parameters of degradation $\mu_{degradation}(x)$ or faults $\mu_{fault}(x)$. The selector of maximum chooses from the fault influence indexes the one that has the maximum value.

The use of the hierarchical representation allows a small amount of "short" fuzzy rules to adequately describe multidimensional dependencies between inputs and outputs.

4. Fuzzy hierarchical Markov state models

A promising approach to constructing intelligent systems of control, diagnosis and monitoring could be the stochastic modelling on the basis of Markov chains combined with the formalized hierarchy theory.

Within a fuzzy hierarchical model, consider fault development processes with the use of Markov chains. Such dynamic models allow to investigate the change of elements' states in time. Fault development can include not only single faults and their combinations, but also sequences (chains) of so-called "consecutive" faults [10, 11].

During FADEC analysis, classification, formalization and representation of processes of condition monitoring and fault diagnosis for the main subsystems of gas turbines (control, monitoring, fuel supply etc.) is carried out. These processes are represented in the form of Markov chains which allow to analyze the state dynamics of the power-plant.

The transition probability matrix of a Markov chain for modeling faults and their consequences, has a universal structure for all levels of system decomposition (Figure 8):

- system as a whole (power plant);
- constructon units;
- elements.

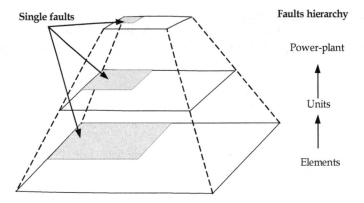

Figure 8. Hierarchical structure of Markov model of fault

The hierarchical Markov model is built in the generalized state space where physical parameters and binary fault flags are used for the estimation of a state vector of the element, unit and power-plant. The state vector includes three parameters { operational, degradation, fault } which allow to track the fault development and degradation process of the system. During FADEC diagnosis, the area of single faults is mostly considered. The proposed Markov model enables to present the system with multiple faults and their sequences. The top state level of a system reflects in the aggregated form the information on faults at the lower state levels.

The elements' state at the levels of the hierarchy depends on the previous values of state parameters of the elements, values of membership functions and fault influence indexes.

For the estimation of transition probabilities between the states of a Markov chain, it is required to calculate relative frequencies of events such as $S_i \rightarrow S_j$ for a given interval of time. In particular, at the top level, the number of events during one flight (Figure 9) can be of interest.

$P_{ij} = \mathrm{Prob}\{S_i \rightarrow S_j\}$	1	2	3
1. Operational state	P_{11}	P_{12}	P_{13}
2. Degradation	P_{21}	P_{22}	P_{23}
3. Fault leading to engine stop	P_{31}	P_{32}	P_{33}

Engine restart in flight

Figure 9. Transition probability matrix of power-plant during one flight

The most important events during flight are the emergency turning off of the engine stop (shutdown) and the possibility of its restart. For the probability estimation of such events, it is required to use statistics on all park of the same type engines. For realization of such estimation methods, it is required that flights information was available on each plane and power-plant. Such information should be gathered and stored in a uniform format and should be available for processing. Modern information technologies open possibilities for such research. To analyze fault development processes of one FADEC, it is possible to use results of the automated tests at the hardware-in-the-loop test bed with modeling of various faults and their combinations. In any case, to receive reliable statistical estimates one needs a representative sample of rather large amount of data.

In the analysis of the Markov model, the relation of the transition probability matrix with state of elements and subsystems at each level of hierarchy is considered. Therefore it is necessary to have the model of the system behavior in various states with various flight condition to guarantee system safety, reliable localization and accommodation of faults.

As the basic mathematical model of the controlled plant, the description in the state space is considered in the form of stochastic difference equations:

$$X(t+1) = \mathbf{A}X(t) + \mathbf{B}U(t) + \mathbf{F}\xi(t), \tag{1}$$

where $X \in R^s$ is the s-dimensioned state vector; $U \in R^s$ is the s-dimensioned control vector; A, B and F are ($n \times n$), ($n \times s$) and ($n \times r$) matrices accordingly; $\xi \in R^s$ is the vector of independent random variables. Thus, the dynamic object described by this finite-difference equation, with input coordinate (control variable) U and output coordinate (state variable) X, in the closed scheme of the automatic control system is the controlled Markov process [12, 13].

The level of quantisation allows the Markov process to be converted into the Markov chain. Provided $\xi(t)$ is a stationary process, the Markov chain will be homogeneous. Such chain is described by the means of the stochastic transition probability matrix P with the dimensions ($m \times m$), where m is the number of the chain states. Each element of the matrix P_{ij} represents the probability of the system transition from the condition X_i into the condition X_j during the time interval ΔT:

$$P_{ij} = \mathrm{Prob}\{X(t) = X_i, \ X(t+1) = X_j\}, \quad \forall n \in N,$$

$$X_i \in \left[x_i - \frac{\Delta x}{2}; x_i + \frac{\Delta x}{2}\right]; \ \sum_{j=1}^{m} P_{ij} = 1, \ i = \overline{1, m}. \tag{2}$$

Condition (2) means that the matrix P should be stochastic and define the full system of events. The sum of elements in each row of the stochastic matrix should equal 1.

The size of the matrix P is defined by the prior information on the order of the object model (1) and the number of the sampling intervals Δx and Δu. The transition probabilities are then estimated as relative frequencies of the corresponding discrete events.

The statistical estimation of the transition probabilities for the controlled Markov chain is performed as the calculation of the frequencies for the corresponding events during observation and the subsequent calculation of the elements of matrix P using the formula:

$$P_{ijk} = \frac{N_{ijk}}{\sum\limits_{j=1}^{m} N_{ijk}} \tag{3}$$

where the numerator N_{ijk} is the number of the following events: $\{X(t_n) = X_i, \ X(t_{n+1}) = X_j, \ U(t_n) = U_k\}$, and the denominator corresponds to the number of events such as $\{X(t_n)=X_i,\ U(t_n)=U_k\}$. Thus, for any combination of state X_i and control U_k, a full system of events will consist of the set of the state transitions X_j.

The normalisation of Equation (3) makes matrix P stochastic. As a result, the set of probabilities in each row P_{ij} describes the full system of events for which the sum of probabilities is equal to unit:

$$\sum_{j=1}^{m} P_{ijk} = 1.$$

The estimation of a transition probability matrix of the Markov model consists of creation of multidimensional histograms which represent an estimate of joint distribution [14, 15].

The use of the hierarchical Markov model allows to "compress" information which has been recorded during one flight, and to present it in a more compact form. In this case, the possibility of analysis and forecast of dynamics of degradation degree (Figure 10) opens. It is possible to analyze the state dynamics of elements and functions at each level of hierarchy in time for decision-making support.

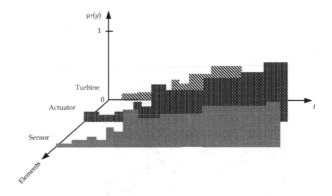

Figure 10. Dynamics of state parameters during flight

The analysis of fault information and state change can be carried out over flight data for the whole duration of maintenance and the whole "fleet" of engines and their systems (Figure 11). Such analysis will assist to increase efficiency for processes of experimental maintenance development and monitoring system support.

Given statistics on all park of engines within several years, it is possible to build empirical estimates of probabilities of the first and second type errors.

Thus, possibilities of application of hierarchical Markov models for the gas turbine and its FADEC for compact representation of information on flight and for the assessment of "sensitivity" of the monitoring system according to actual data are considered. The levels of hierarchy differ with the ways of introducing redundancy and realization of system safety with use of intellectual algorithms of control and diagnosis. Each higher level of hierarchy has greater "intelligence" and is designed independently in the assumption of ideal system stability of the lower level.

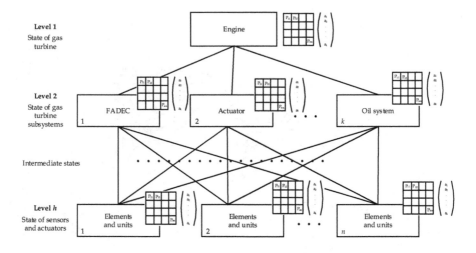

Figure 11. Hierarchical fuzzy Markov model of gas turbine states

Consider an example. In Figure 12, the FADEC state estimation with faults is presented on the basis of the degradation degree of the elements.

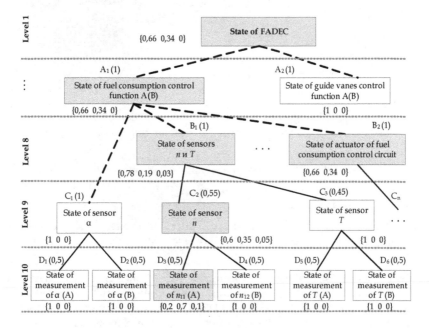

Figure 12. Example of hierarchical estimation of state parameters

At the 10th level the BMS detected a fault of measurement in the form of break of the first coil of parameter n_{11} (shaft speed sensor). On the basis of the fuzzy rule $R^{(2)}$, the parameters of measurement state of n_1 in the channel A are characterized by the following three values

$$\begin{bmatrix} \mu_{T_1}(n_{11})=0,2 \\ \mu_{T_2}(n_{11})=0,7 \\ \mu_{T_3}(n_{11})=0,1 \end{bmatrix}.$$

$$R^{(2)} : IF\,(n_{11} = A_2\; AND\; n_{12} = A_1)\; THEN\; y = T_2 \Rightarrow$$
$$[\mu_{T_1}(n_1)=0,2;\quad \mu_{T_2}(n_1)=0,7;\quad \mu_{T_3}(n_1)=0,1].$$

The measurement state in the channel B is defined as $\begin{bmatrix} \mu_{T_1}(n_{12})=1 \\ \mu_{T_2}(n_{12})=0 \\ \mu_{T_3}(n_{12})=0 \end{bmatrix}$, because no faults were de-

tected. At the 9th level, the sensor state of the n is obtained using the multiplication of the vector of state parameters and faults influence indexes:

$$\begin{bmatrix} S(n_{11},n_{12})_o \\ S(n_{11},n_{12})_d \\ S(n_{11},n_{12})_f \end{bmatrix} = \begin{bmatrix} \mu_{T_1}(n_1) = 0,2 \\ \mu_{T_2}(n_1) = 0,7 \\ \mu_{T_3}(n_1) = 0,1 \end{bmatrix} \begin{bmatrix} \mu_{T_1}(T_4) = 1 \\ \mu_{T_2}(T_4) = 0 \\ \mu_{T_3}(T_4) = 0 \end{bmatrix} \times [0,5; \ 0,5] =$$

$$= \begin{bmatrix} (0,2 \times 0,5) + (1 \times 0,5) \\ (0,7 \times 0,5) + (0 \times 0,5) \\ (0,1 \times 0,5) + (0 \times 0,5) \end{bmatrix} = \begin{bmatrix} 0,60 \\ 0,35 \\ 0,05 \end{bmatrix}.$$

(ID1)

The state of an element of a higher level is calculated by multiplication of the current state to fault influence indexes of the fault elements. The state of both measurement channels of temperature T is "operational", therefore, the sensor T state equals $\begin{bmatrix} S(T)_o = 1 \\ S(T)_d = 0 \\ S(T)_f = 0 \end{bmatrix}$ The sensor of fuel feed α is also good working.

At the 8th level the state of two sensors n and T after similar calculations becomes equal { 0,78; 0,19; 0,03 } that indicated the system degradation in the part of control of fuel consumption.

For the estimation of a state of the fuel consumption control function, the "OR" operation is also used. The state of the actuator of fuel consumption control circuit is characterized by the parameters { 0,66; 0,34; 0 }. The state of FADEC is characterized by the fault of fuel consumption control function or the state of the guide vanes control function. Using the operation "OR", the state of FADEC is detected as { 0,66; 0,34; 0 }. In this example, the whole system is considered to be operational, whereas partial degradation is observed, which is not influencing the system operability.

Thus, the technique of state parameters determination for FADEC and its systems on the basis of fuzzy logic and Markov chains is proposed. This technique can be used during flight or in maintenance on the ground.

At the present time a necessary condition for realization of intellectual algorithms is the complete development of the distributed intellectual control models focused on control optimization, forecasting and system safety [16, 17]. In Figure 13, the scheme of the distributed FADEC is shown.

Thus, in each sensor or actuator, it is necessary to have physically built-in control system (or function) to form and monitor the fault signals in the unit, communication lines, cooperating sensors, indication devices and systems [18]. The use of the built-in monitoring control systems working in real time allows to obtain a number of additional possibilities for improving control quality and system operational characteristics as followings:

• emergency states detection of the control object and system;

• fault detection of elements of the control object;

- state diagnosis and parametrical degradation of the object.

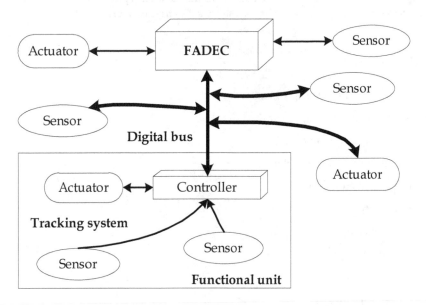

Figure 13. Distributed architecture of FADEC

5. Conclusion

In this chapter, the hierarchical fuzzy Markov modeling of fault developments processes has been proposed for the analysis of an airplane system safety. The hierarchical model integrates functional, physical structure of gas turbine and its FADEC elements and units, the tree of states, a tree of fault influence indexes. This model allows to decompose the power-plant for a quantitative estimates of degradation state and gradual faults. The analysis of hierarchies allows to utilize the state model on the basis of fault development processes which estimates the power-plant state at each level of hierarchy. Furthermore, the technique of determination of state parameters of the gas turbine and its systems on the basis of fuzzy logic is presented. The state of each element, unit and system is represented in the form of a vector with parameters { operational, degradation, faults }. The use of the proposed indicator "degradation degree" allows to obtain an objective quantitative estimate of the current state which can be used as, the "distance" to a critical situation and the reserve of time for decision-making in-flight. This indicator is defined on the basis of the discrete-ordered scale and fault influence indexes that allows to determine about 30 % of gradual faults in gas turbine and its systems at the stage of fault development. The examples of fuzzy rules on the basis of expert knowledge are given, whereas fuzzy logic is used for interpolation.

The application of hierarchical Markov models for the analysis of experimental data is also considered for control system development: as the compact representation of information of a system state change during flight, the estimation of transition probabilities.

Nomenclature

FADEC – Full Authority Digital Engine Control.

BMS – built-in monitoring system.

ω – weighting coefficient of fault.

S – state (condition) of gas turbines.

μ – membership function.

P – probability.

X – state variable.

U – control variable.

Acknowledgements

The work was supported by the grants from the Russian Foundation for Basic Research (RFFI) №12-08-31279, №12-08-97027 and the Ministry of Education and Science of the Russian Federation.

Author details

G. G. Kulikov, V. Yu. Arkov and A.I. Abdulnagimov*

*Address all correspondence to: gennadyg_98@yahoo.com; abdulnagimov@gmail.com

Automated Control and Management Systems Department, Ufa State Aviation Technical University, Ufa, Russia

References

[1] Kulikov GG. Principles of design of digital control systems for aero engines. In Cherkasov BA. (ed.), Control and automatics of jet engines. Mashinostroyeniye, Moscow; 1988. p253-274.

[2] Arkov VY, Kulikov GG, Breikin TV. Life cycle support for dynamic modelling of gas turbines. Prepr. 15th Triennial IFAC World Congress, Barcelona, Spain; 2002. p2135-2140.

[3] Kuo BC. Automatic control systems. Prentice-Hall: Englewood Cliffs; 1995.

[4] Saaty TL. Analytic Hierarchy Process: Planning, Priority Setting, Resource Allocation. McGraw-Hill, New York, London; 1980.

[5] Saaty TL, Vargas LG. Models, Methods, Concepts & Applications of the Analytic Hierarchy Process. Kluwer Academic Publisher; 2001.

[6] SAE ARP 4761. Guidelines and Methods for Conducting the Safety Assessment Process on Civil Airborne Systems And Equipment. Aerospace recommended practice; 1996.

[7] Arkov V, Evans DC, Fleming PJ, et al. System identification strategies applied to aircraft gas turbine engines. Proc. 14th Triennal IFAC World Congress; 1999. p145-152.

[8] Kulikov G, Breikin T, Arkov V and Fleming P. Real-time simulation of aviation engines for FADEC test-beds. Proc. Int. Gas Turbine Congress, Kobe, Japan; 1999. p949-952.

[9] Zadeh LA. "Fuzzy algorithms," Information and Control 1968;12(2): 94-102.

[10] Kulikov GG, Fleming PJ, Breikin TV, Arkov VY. Markov modelling of complex dynamic systems: identification, simulation, and condition monitoring with example of digital automatic control system of gas turbine engine. USATU, Ufa; 1998.

[11] Breikin TV, Arkov VY, Kulikov GG. On stochastic system identification: Markov models approach. Proc 2nd Asian Control Conf ASCC'97; 1997. p775-778.

[12] Breikin TV, Arkov VY, Kulikov GG. Application of Markov chains to identification of gas turbine engine dynamic models. International Journal of Systems Science 2006; 37(3) 197-205.

[13] Kulikov G, Arkov V, Lyantsev O, et al. Dynamic modelling of gas turbines: identification, simulation, condition monitoring, and optimal control. Springer, London, New York; 2004.

[14] Kulikov G, Arkov V, Abdulnagimov A. Markov modelling for energy efficient control of gas turbine power plant. Proc. IFAC Conf. on Control Methodologies and Technology for Energy Efficiency, CMTEE-2010, Faro, Portugal; 2010. http://www.ifac-papersonline.net/Detailed/42981.html

[15] Arkov V, Kulikov G, Fatikov V, et al. Intelligent control and monitoring unit and its investigation using Markov modelling. Proc. IFAC Int. Conf. on Intelligent Control Systems and Signal Processing ICONS-2003, Faro, Portugal; 2003. p489-493

[16] Vasilyev VI, Ilyasov BG, Valeyev SS. Intelligent Control Systems for gas Turbine engines. Proc. of the Second Scientific Technical Seminar on GT Engines, Turkey, Istanbul; 1996. p71-78.

[17] Culley D, Thomas R and Saus J. Concepts for Distributed Engine Control, AIAA-2007-5709, 43rd AIAA/ASME/SAE/ASEE Joint Propulsion Conference and Exhibit, Cincinnati, Ohio, July 8-11, 2007.

[18] Kulikov GG., Arkov VYu., Abdulnagimov AI. Hierarhical Fuzzy Markov Modelling for System Safety of Power-plant : Proc. of 4th International Symposium on Jet Propulsion and Power Engineering, September 10-12, 2012, Xi'an, China: Northwestern Polytechnical University Press; 2012. p. 589-594.

Chapter 9

Engine Condition Monitoring and Diagnostics

Anastassios G. Stamatis

Additional information is available at the end of the chapter

1. Introduction

Any engine exhibits the effects of wear and tear over time. Several mechanisms cause the degradation and potential failures of gas turbines such as dirt build-up, fouling, erosion, oxidation, corrosion, foreign object damage, worn bearings, worn seals, excessive blade tip clearances, burned or warped turbine vanes or blades, partially or wholly missing blades or vanes, plugged fuel nozzles, cracked and warped combustors, or a cracked rotor disc or blade.

Fouling is caused by liquid or solid particles accumulated to airfoils and annulus surfaces. Deposits consist of varying amounts of moisture, oil, soot, water-soluble constituents, insoluble dirt, and corrosion products of the compressor blades material whish are held together by moisture and oil. The result is a build-up of material that causes increased surface roughness and to some degree changes the shape of the airfoil. Hot corrosion is the loss or deterioration of material from flow path components caused by chemical reactions between the component and certain contaminants, such as salts (for example sodium and potassium), mineral acids or reactive gases (such as hydrogen sulfide or sulfur oxides). Corrosion is caused by noxious fumes or ash-forming substances present in the fuel such as aluminum, calcium, iron, nickel, potassium, sodium, silicon, magnesium. Corrosion increases surface roughness and causes pitting. Erosion is the abrasive removal of material from the flow path by hard or incompressible particles impinging on flow surfaces. Damage may also be caused by foreign objects striking the flow path components (Figure. 1a). Foreign Object Damage (FOD) is defined as material (nuts, bolts, ice, birds, etc.) ingested into the engine from outside the engine envelope. Domestic Object Damage (DOD) is defined as objects from any other part of the engine itself.

Different causes and mechanisms of performance deterioration of jet engines are reviewed in [1]. Degradation in both land and aero gas turbines is also reviewed by Kurz and Brun

[2], who pointed out differences in mechanisms for the two types. Industrial gas turbine deterioration has been discussed by Diakunchak [3].

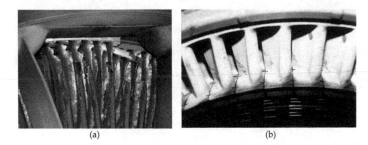

(a)   (b)

Figure 1. (a) FOD effects, (b) Turbine nozzles with deposits.

Three major effects determine the performance deterioration of the gas turbine compressor due to fouling: Increased tip clearances, changes in airfoil geometry, and changes in airfoil surface quality. In compressors, erosion increases tip clearance, shortens blade chords, increases pressure surface roughness, blunts the leading edge, and sharpens the trailing edge. Turbine blade oxidation, corrosion and erosion are normally longtime processes with material losses occurring slowly over a period of time. However, damage resulting from impact by a foreign object is usually sudden. Impact damage to the turbine blades and vanes will result in parameter changes similar to severe erosion or corrosion. Corrosion, erosion, oxidation or impact damage increases the area size of the turbine nozzle. When crude oil is burned in the GT the hot end is subjected to additional harmful deposits, including salt deposits originating in the inlet or from fuel additives. As hot combustion products pass through the first stage nozzle, they experience a drop in static temperature and some ashes may be deposited on the nozzle blades decreasing the nozzle area (Figure 1b). The combustion system is not likely to be the direct cause for performance deterioration. The combustion efficiency will usually not decrease, except for severe cases of combustor distress. However, plugged nozzles and/or combustor and transition piece failures will always result in distorted exhaust gas temperature patterns. This is a result of the swirl effect through the turbine from the combustor to the exhaust gas temperature-measuring plane. Distortion in the temperature pattern or temperature profile not only affects combustor performance but can have a far reaching impact as local temperature peaks can damage the turbine section.

All the above causes and effects may be considered as faults. Generally speaking, fault is a condition of a machine linked to a change of the form of its parts and of its way of operation, from what the machine was originally designed for and was achieved during its initial operation. In this respect a fault manifests itself by a change of geometrical characteristics or/and integrity of the material of parts of an engine. Change in geometry is inevitably linked to common experience faults, as for example when a part is broken, or deformed. Typical integrity fault is the occurrence of cracks inside the material, which are not associated to any geometrical change but can nevertheless result into catastrophic consequences. Some of the faults will become evident as vibration increases or by a change in lubrication oil temperature. However, some serious faults can be detected only through gas path analysis. The gas path, in its simplest form, consists of the compressors, combustor, and turbines.

Diagnosis of a mechanical condition is the ability to infer about the condition of parts of the engine, without dismantling the engine or getting direct access to these parts, but only from observations of information coming to the engine exterior. The field of engineering science covering the techniques for achieving a diagnosis is called diagnostics. The aim of diagnostics is to detect the presence and identify the kind of faults appearing in a engine. Diagnostics does not require that the engine is either stopped or disassembled. Information is gathered while the engine is in operation. This is vital for engines in the process industry or energy production, as they must run without interruption for long time intervals. Detection of an incipient failure in a jet engine leads to taking action necessary to prevent a catastrophic failure which might follow.

In order to establish the possibility of diagnosing engine condition a correspondence of this condition to the values of the measured quantities should be known. In general terms, this correspondence is intrinsically established through the physical laws governing the operation of the machine. The behavior of any relevant physical quantity is linked through these laws to the detailed geometry of the machine and the kind of phenomena taking place in it. If we consider a machine using a fluid as a working medium, the variation of the flow quantities at one particular location in the machine is determined, via the laws of fluid mechanics, from the geometry of the solid boundaries and the physical properties of the fluid. A change in geometry will then reflect on the values of the flow quantities and could be calculated by application of the relevant physical laws. If suitable quantities are measured, they reflect changes in geometry or material and can therefore be used to indicate the presence of a fault. It is obvious that according to the change occurring in an operating machine, different quantities will be influenced. For example, the operation of rotating components is always linked to the exertion of periodic forces, with a frequency which is usually a multiple of the frequency of rotation. In this respect, the quantities characterizing a vibration are suitable for diagnostic purposes. On the other hand, severe corrosion, as it changes turbine airfoil geometry, is detectable through gas path analysis.

Many techniques for inferring engine status or change in engine condition have been proposed and/or applied to various engine configurations with varying success. Some of them (e.g. Vibration monitoring, Trending Analysis) are parts of computer-controlled data-acquisition systems that permit the on-line acquisition and reduction of a very large amount of performance information. While fault detection or general deterioration could be based on immediate observation of reduced measurable quantities, such observation is not, generally, adequate. It should also be noted that a change in any measured parameter does not necessarily indicate a particular independent parameter fault. For example, a change in compressor discharge pressure (CDP) does not necessarily indicate a dirty compressor. The change could also be due to a combined compressor and turbine fault or to a turbine fault alone. In order to have access to the variables, which possess diagnostic information (such as component efficiencies) modeling of an engine is essential. Thermodynamic (Gas Path) analysis methods employ engine models to

process measurement data, in order to diagnose changes in component performance which
may be linked to degradation, aging, or incipient failure.

2. Gas path analysis

An engine may be viewed as a system, whose operating point is defined by means of a set of
variables, denoted as **u**. The operation of each component follows predictable thermodynamic laws. Therefore, each component will behave in a predictable manner when operating
under a given set of conditions. The health condition of its components is assumed to be
represented through the values of a set of appropriate "health" parameters such as efficiencies and flow capacities, contained in a vector **f**. The system is observed through measured
variables, such as speeds, pressures, temperatures, contained in a vector **y**. When the engine
operates at a certain operating point measured quantities are produced for given values of
health parameters. The operating engine establishes a relationship between these parameters, which can be expressed though a functional relation:

$$\mathbf{y} = \mathbf{F}(\mathbf{u}, \mathbf{f}) \tag{1}$$

A computer model materializing this relation can reproduce the values of any thermodynamic quantity measured along the engine gas path. It is interesting to note that by assigning appropriate values to the components of vector **f**, the effect of engine component faults
or deterioration on measured quantities can be reproduced.

The problem of diagnostics (Figure 2) is to seek a solution to the inverse problem, namely to
determine the values of the estimated health parameters $\hat{f}$ from a given set of measurements using a diagnostic method (*DM*). Particular faults can then be detected if deviations
of health parameters from the reference state are observed.

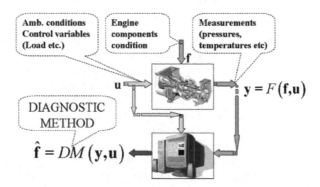

Figure 2. Gas Path Analysis diagnostics formulation.

Many variants of Gas Path Analysis based diagnosis with different features and complexity have been developed and reported in the open literature. Extensive reviews of existing methods provided by Li [4], and Marinai et al. [5].

Generally speaking any GPA method at least consists of the following elements:

• Measured data

• A data processing model relating measured data with health parameters

• A diagnostic decision making procedure.

The data used can be taken in steady state or transient operation. The model could be a physical one representing the aerothermodynamic processes taking place in the engine components and the mechanical coupling between them or a black box mathematical model relating data with health parameters. The diagnostic decision making procedure may be a conventional pattern recognition technique applied to health parameter space or an artificial intelligence based expert system.

Accordingly the proposed methods are classified on the basis of the kind of the comprising elements as: Steady state or Transient, Physical or Mathematical, Conventional or Artificial intelligence method.

3. Physical models based GPA

3.1. Linear methods

In linear gas path analysis, the health parameters are represented as the unknown "deltas" of component performance parameters (typically efficiency and mass flow capacity). They are related to known measurement "deltas" through relations produced by linearization of the general nonlinear thermodynamic relations, assuming small deviations. [6].The classical linear approach is formulated as follows: For a given operating point u the measurement values depend only on the health condition of engine components. After linearization and taking into account measurement uncertainty (by adding a noise vector **v** with zero mean and known covariance R), the typical GPA equations take the form:

$$\Delta y = C \cdot \Delta f + v \tag{2}$$

where Δ is called delta and represents percentage deviation from a reference value(when the engine is in intact condition) and C the well-known influence coefficient matrix. Estimation of health parameters is obtained from the relations

$$\Delta \hat{f} = S^{-1} \cdot C^T \cdot R^{-1} \cdot \Delta y \tag{3}$$

$$S = M^{-1} + C^T \cdot R^{-1} \cdot C \tag{4}$$

where M represents known statistics for the deviation of health parameters.

Although the formulation for classical GPA has proven to be successful for practical purposes and existing commercial systems ([7, 8]) are based on it, identifiability problems exist due to limited instrumentation. Sufficient engine health assessment requires at least the estimation of the parameters associated with the main engine components. Considering an existing engine, a typical situation is characterized by the fact that the number of available sensors is smaller than the number of parameters to be calculated. Accordingly, all the initially implemented methods were compelled to adopt various assumptions. Most of the methods use a priori information about the statistics of the calculated parameters introducing thus bias in the estimation. In that case, inversion of matrix S is only possible when it is dominated by M. The main drawback is the effect discussed by Doel [9]. The algorithm tends to "smear" the fault over many components.

3.1.1. Multi operating point GPA

GPA Multi Operating Point Analysis (MOPA) methods have been developed trying to exploit information provided by the existing sensors when different operating points are considered. The origin for the multi operating point analysis (MOPA) methods was the Discrete Operating point GPA, introduced in [10].The method, based on information given by existing sensors when different operating points are considered, improved significantly the diagnostic effectiveness. The implementation of the method was an extension of the classical linear gas path analysis. MOPA methods though do not use a priori statistics for the parameters rely on the questionable assumption of non-varying health parameters. Other research groups applied the same principle for the nonlinear case, [11-13].

The linear implementation for the MOPA approach using NOP operating points is given by Eqs. (5)-(9).

$$\Delta y_k = C_k \cdot \Delta f \quad k = 1, NOP \tag{5}$$

$$C_k = \left[c_{ij,k} \right] \tag{6}$$

$$c_{ij,k} = \left(\partial \Delta y_i / \partial \Delta f_j \right)_k \tag{7}$$

$$\Delta \hat{f} = P^{-1} \cdot \sum_{k=1}^{NOP} \left(C_k^T \cdot R_k^{-1} \cdot \Delta y_k \right) \tag{8}$$

$$P = \sum_{k=1}^{NOP} \left(C_k^T \cdot R_k^{-1} \cdot C_k \right) \tag{9}$$

The so called information matrix P is crucial in the sense that its condition determines the diagnostic effectiveness. The condition of the matrix is represented by its condition number. Investigations concerning effects of both the number of operating points used and the 'distance' of the operating points on information matrices have been reported ([14]-[15]). Additional details on assessing identifiability in multipoint gas turbine estimation problems are given in [15]. Although all the works implementing the multipoint approach agree that the idea more or less improves the diagnostic effectiveness, there are also results (see [16]), indicating that the theoretically attainable multi-point improvements are difficult to realize in practical engine applications.

In order to understand the reasons for potential problems concerning diagnosis using a multipoint approach it is necessary to examine the underlying assumptions of the method. The main assumption of the method is that the 'deltas' concerning the health parameters remain constant with regard to change in operating conditions. This assumption is obviously true for some parameters (for example the parameter expressing the effective turbine area or the area of non-variable nozzle jet engine), but there are indications that for other parameters this is a week assumption. Several works ([3], [17]), have provided evidence that when deterioration is present, the deviations of parameters such as flow compressor capacity and efficiency change with the operating point. In fact different working-point means different aerodynamic conditions and, in this sense, efficiencies and flow capacities deltas can significantly vary with the operating condition. The resulting diagnosis risk is not only to imprecisely calculate the engine new state after some deterioration but even more to indicate as responsible for the fault the wrong component(s).

Recently a new variant of GPA method named Artificial Multi Operating Point Analysis (AMOPA) has been proposed [18]. The new method uses existing sensor information produced when artificial operating points are used close to an initial operating point by using different parameters for each operating point definition. Therefore the assumption that the 'deltas' of the health parameters remain constant is reasonable. The method proved to be capable of both isolating and identifying the fault in individual components.

3.2. Nonlinear methods

In nonlinear methods, the full thermodynamic equations are treated directly without simplification. An example of such a method, the method of adaptive modeling introduced by Stamatis et al. [19], uses component maps "modification factors" as health parameters and solves for them through an optimization procedure applied to a function based on differences of the predicted and measured values. Variants of the nonlinear GPA have been proposed (see [20-22]), the main differences being the objective function formulation as well as the method used for the optimization. The more general objective function (OF) to be minimized was proposed in ref. [23]:

$$OF = \sum_{i=1}^{m} \left[\frac{y_i^{calc}(\mathbf{f}) - y_i}{y_i \sigma_{Y_i}} \right]^2 + C_A \cdot \sum_{j=1}^{n} \left| \frac{f_j - f_j^r}{f_j^r \sigma_{f_j}} \right| + C_S \cdot \sum_{j=1}^{n} \left[\frac{f_j - f_j^r}{f_j^r \sigma_{f_j}} \right]^2 \qquad (10)$$

where n and m the dimensionalities of $\mathbf{f}$ and $\mathbf{y}$ correspondingly. The first term express the fact that the health parameters under estimation $\mathbf{f}$ must be such that the values of measured quantities $\mathbf{y}$ are reproduced as accurately as possible. The second and third terms ensure that the values of health parameters cannot be significant different from their reference, a fact resulting from experience. It is the addition of these terms that allows the derivation of a solution for $\mathbf{f}$, even when a smaller number of measurements is available. All *deltas* are weighted by the inverse of the standard deviation of the corresponding quantity. Weight factors C_A, C_S are also included, for the possibility to change the relative importance of the two groups of terms. The reference values $\mathbf{f}^r$ of the health parameters can be chosen to represent a 'best' guess of the values to be determined. From studies in estimation theory, it has been found that it is useful to include in the objective function a term of sum of absolute values, since this term may improve the numerical behavior of the estimation procedure by increasing its robustness (see [24]).

The way of determining the vector $\mathbf{f}$ for minimization of this function can take advantage of the physical characteristics of the problem to be solved. For example the fact that deviation of component efficiencies should not be positive could be formulated as a constraint in the optimization. In the case of slow deterioration tracking, the reference values can be chosen to vary slowly with time while a filtering procedure can be applied, taking advantage of the regular variation of component deviations, as described in [25]. For the case of individual component faults the fault usually affects one or two neighboring components.

All these features should be taken into account when formulating the diagnostic algorithm. The solution is obtained with the interaction of a non-linear engine performance model and an optimisation algorithm, as shown in figure 3.

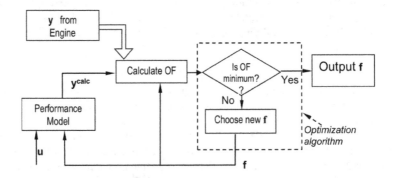

Figure 3. Schematic representation of nonlinear diagnostic procedure

The methodology for diagnosing single component faults using the above procedure is based on the following reasoning. Since measurement data are noisy, the estimations based on a single data set differ from the actual values due to noise propagation. They can be improved when more than one measurement data sets are available. In such a case a solution is obtained for each individual data set. A series of values for each health parameter f_j becomes thus available. The mean value and standard deviation of the percentage change from reference Δf_j are then calculated. A criterion then is proposed for isolating the parameters of the components that are faulty, with the aid of a parameter, which we call diagnostic index. We define as diagnostic index the ratio of the absolute mean value to the standard deviation for each estimated health parameter.

$$DI_j = \frac{\left|\overline{\Delta f_j}\right|}{\sigma_{f_j}}$$

(11)

Health parameters exhibiting small deviations from reference state or parameters with large standard deviations (large uncertainty on derived estimations) will have small values for diagnostic index. On the other hand, health parameters with large mean value or small standard deviation (small uncertainty on derived estimations) will present large values for diagnostic index. It is thus expected that the health parameters, which deviate due to fault occurrence will be those with the largest value of the diagnostic index. Thus we identify as faulty the component containing the parameter with the largest diagnostic index. This stage is called fault localization.

After the detection of a faulty component, a more accurate estimation of fault magnitude can be performed. The optimization problem is solved again by keeping as unknowns only the health parameters of the component found faulty. C_A, C_S are zeroed, to avoid biases imposed by the corresponding terms. (Note that with much fewer unknowns a unique solution can be derived by minimizing differences only from measurements namely the first term of the objective function eq (10).) After performing a series of estimations with this formulation from the available data sets, the average values of the obtained parameters are kept as the estimations for the fault magnitude.

Nonlinear GPA methods have proved accurate and robust provided that appropriate measured variables and estimated health parameters have been selected. This is not a trivial problem as explained in the following.

3.3. Sensors and health parameters selection

When application of a GPA technique is envisaged on an engine, the existence of certain restrictions is recognized. Considering an existing engine, there is always a given set of available measurements. Addition of instrumentation can be difficult or even impossible. It is therefore important to have the possibility to adopt a convenient formulation of a method, so that an optimal use of the existing measurements is achieved. On the other hand, when a new engine is designed, or when an intervention to instrument an engine is performed, it is desirable to define an optimum combination of sensors to be installed.

The problems that may be faced in such a situation can be summarized as follows: When a given set of measured quantities is provided, what is the optimum set of health parameters? The particular problem is to define the best possible parameters for a given measurement possibility. This is a problem faced usually by the engine user, who has very few or no possibilities of intervening and adding measurements on the engine. When the decision for instrumenting an engine has to be taken, both the manufacturer and the user are faced with the inverse problem: (a) The user wants to know the optimum set of measuring instruments to be added in order to provide enough information for a required level of resolution. (b) The manufacturer wants to decide which instruments will accompany the engine, in order to ensure a good capability of in-service monitoring.

A systematic study for methods of choice of measurements and parameters in a way optimal as to diagnostic effectiveness was first presented by Stamatis et al. [26]. They introduced criteria for optimal measurement or health parameter selection. We present here the proposed method for measurement selection. Let $f^{(r)}$ be the baseline diagnostic vector corresponding to a healthy engine (typically $f^{(r)} = I$), and $f^{(j)}$ the diagnostic vector resulting when the jth element of $/$ deviates from the baseline value ($f(r)$) by a percentage amount h_j.

$$f^{(j)} = f^{(r)} \cdot + h_j \cdot e^{(j)} \quad j = 1, ..., m \tag{12}$$

hj is a small constant (0.001 $< h_j <$ 0.01). Then, from Eq. (3) we have

$$Y^{(r)} = F\left(f^{(r)}\right) \tag{13}$$

$$Y^{(j)} = F\left(f^{(j)}\right) = F\left(f^{(r)} + h_j \cdot e^{(j)}\right) \tag{14}$$

The sensitivity of each dependent parameter on each individual health index is evaluated as

$$\Delta Y_k^{(j)} = \left(Y_k^{(j)} - Y_k^{(r)}\right) / Y_k^{(r)} \quad k = 1, ..., n \tag{15}$$

We also define an overall sensitivity measure for each parameter with the norm

$$SY_k = \left[\frac{1}{m} \cdot \sum_{j=1}^{m} \left(\Delta Y_k^{(j)}\right)^2\right]^{1/2} \quad k = 1, ..., n \tag{16}$$

So, the problem of selecting the appropriate measurements is expressed mathematically as follows: For a given set of health condition parameters, we must select as measured parameters these parameters giving on the norm of Eq. (16) the m greater values.

In later years more works have appeared, approaching the problem from different points of view, [14, 27].

4. Artificial intelligence GPA methods

4.1. Neural networks

An Artificial Neural Network (ANN) is an information processing paradigm that is inspired by the way biological nervous systems, such as the brain, process information. The key element of this paradigm is the novel structure of the information processing system. It is composed of a large number of highly interconnected processing elements (neurons) working together to solve specific problems. ANNs, like people, learn by example. Learning in biological systems involves adjustments to the synaptic connections that exist between the neurons. This is true of ANNs as well. Two tasks which can be performed by neural nets, which are relevant to the procedure of monitoring and diagnostics of a gas turbine, are: modeling the performance of a gas turbine and detection and classification of faults.

A typical use of a model is to produce reference values for quantities which are monitored. It can be also used for other purposes such as generation of influence coefficient matrices, and sensitivity analyses. ANNs are known to be able to model non-linear systems and therefore can be used for gas turbine performance modeling. A first advantage offered by modeling engine performance through ANN is the much shorter computational time required, once the net is trained and verified, in comparison to any full scale aerothermodynamic model. The latter involves the solution of a set of non-linear equations, which is achieved through iterative schemes, resulting in a number of arithmetic operations significantly larger than those performed by an ANN. A further advantage is related to the possibility of adapting to a particular engine, if data is available. A well-known fact is that for a model to be accurately representing the operation of an engine, it has to be adapted to the particular engine (as discussed, for example, in [19]). A model using ANN provides inherently this possibility, through the way it is being set up. The existence of a learning phase, (called "training" in the ANN terminology) allows the adaptation to a particular engine, if enough data is available.

The second area of possible application, detection and identification of faults, comes from one of the most powerful capabilities of ANN, namely the capability of identifying and classifying patterns. Any method of fault detection and identification uses a set of changes in the values of some parameters, to detect and identify a component malfunction. The task of assigning such sets of changes to machine status is one very much suited to ANN. Neural networks, with their remarkable ability to derive meaning from complicated or imprecise data can be used to extract patterns and detect trends that are too complex to be noticed by either humans or other computer techniques.

There are various neural network models. Among all different neural networks, the back-propagation and the probabilistic neural nets are the architectures, which have mostly been investigated for gas turbine diagnostics. The majority of the researchers refer to performance diagnostics [28, 29], while fewer refer to sensor fault detection and isolation. Kanelopoulos et al. [30] studied the performance of back-propagation (BP) neural nets for both sensor and actual engine component faults for a single shaft industrial gas turbine. The BP neural networks, however, have two main limitations: (1) difficulty of determining the network structure and the number of nodes; (2) slow convergence of the training process.

Probabilistic Neural Networks (PNN), exhibit certain advantages that make them attractive, a significant one being that their particular structure does not require a training procedure, needed for other types of neural networks. The training information is produced during the network set-up and is then embedded in its structure. PNN 'training' can thus be considered to be much faster than for other types of network, such as back-propagation. Additionally, PNNs perform a probabilistic rather than a deterministic diagnosis, something closer to physical reality.

4.1.1. Probabilistic Neural Networks (PNN)

The Probabilistic Neural Network (PNN) is a multi-layer feed forward network. The learning procedure of this network is a supervised learning procedure. During the learning procedure the PNN classifies the training patterns to classes (represented by the output nodes). When an unknown pattern is presented to the PNN, the network estimates the probability that this pattern belongs to each class. The procedure followed and the network itself is briefly described in the following:

Let us suppose that, for training the PNN, we use the group of the m, n-dimensional, training patterns:

$$\mathbf{x}_j = \left\{ a_{1j}, \quad a_{2j}, \quad .... \quad a_{nj} \right\}, \; j = 1,...m \tag{17}$$

The graph of the resulting network is shown in figure 4. The PPN consists of three layers. The n nodes of the first layer represent the n-dimensional input. The m nodes of the second layer (hidden layer) represent the training patterns, while each one of the k nodes of the third (output) layer represents a class to which a pattern can be classified into.

Every node of the input layer of the PPN is linked to every node of the hidden layer. Each node of the hidden layer (representing a training pattern) is linked only to the node of the output layer that represents the class where the training pattern 'belongs'.

When a pattern $\mathbf{x} \in^m$ is given as an input to the network, the output is the probability density functions: $P(S_i \mid \mathbf{x})$, i=1,...,k.

If we assume that the probability density functions, $P(\mathbf{x} \mid S_i)$, are Gaussian, we have:

$$P(S_i \mid \mathbf{x}) = \frac{P(S_i)}{P(\mathbf{x}) \cdot (2\pi)^{m/2} \cdot \sigma_i^m \cdot |S_i|} \cdot \sum_{j=1}^{n_i} \exp\left[\frac{-(\mathbf{x} - \mathbf{x}_j^{(i)})^T (\mathbf{x} - \mathbf{x}_j^{(i)})}{2\sigma_i^2}\right] \tag{18}$$

where, $x_j^{(i)}$ is the j-th pattern of the training set of patterns that 'belong' to class i, $|S_i|=n_i$ is the number of the training patterns that 'belong' to class i, σ_i is a smoothing parameter, $P(S_i)$ is the 'a priori' probability of class S_i, and $P(x)$ a normalization factor representing the 'a priori' probability of pattern x, which is constant assuming mutually exclusive classes, covering all possible situations.

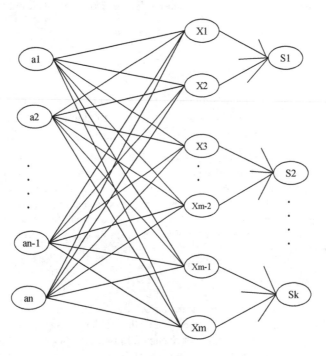

Figure 4. The general structure of the Probabilistic Neural Network.

For example if it is considered that the 'a priori' probability is equal for all classes,

$$P(S_i) = \frac{1}{k}, \; i = 1, \ldots, k \tag{19}$$

During the training of the PNN, we provide the training patterns and the classes they belong to. From this information the number of nodes of each layer, as well as the links of the network with the related weights, are specified.

The weight of the link from node j1 of the input layer to node j2 of the hidden layer is:

$$w^{(1)}_{j1,j2} = a_{j1j2} \tag{20}$$

while, the weight of the link from node X_i of the hidden layer to node S_j of the hidden layer, is:

$$w^{(2)}_{X_i,S_j} = \frac{1}{2 \cdot \sigma_j^2} \tag{21}$$

where, σ_j is the smoothing parameter of class j, represented by node S_j of the output layer of the network. During the testing of the network, the probability density functions for each class are calculated, using equation (18).

Comparative and parametric investigations of the diagnostic ability of PNN on turbofan engines have been carried out in [31]. The work has also provided some general information about PNN diagnostic ability. The use of probabilistic neural networks for sensor fault detection and estimation of the sensor bias has been demonstrated in [32]. The technique proposed was shown to provide a powerful sensor validation tool, for cases where a rather limited number of measuring sensors is available, such as when data from an engine onboard an aircraft are available.

4.2. Expert systems

In contrast to neural networks, which learn knowledge by training on observed data with known inputs and outputs, Expert systems(ES) utilize domain expert knowledge in a computer program with an automated inference engine to perform reasoning for problem solving. Three main reasoning methods for ES used in the area of engine diagnostics are rule-based reasoning, case-based reasoning and model-based reasoning. In condition monitoring practice, knowledge from domain specific experts is usually inexact and reasoning on knowledge is often imprecise. Therefore, measures of the uncertainties in knowledge and reasoning are required for ES to provide more robust problem solving. Commonly used uncertainty measures are probability, fuzzy member functions in fuzzy logic theory and belief functions in belief networks theory. An expert system dealing with uncertainty and proved to be very efficient in fault diagnosis is described below.

4.2.1. Bayesian Belief Network (BBN)

BBN is a probabilistic expert system, graphically represented by a set of 'nodes' and a set of 'links' connecting them. The topological features of a BBN that must be fully specified in order the network to be complete are the following: Nodes express the parameters of the represented domain. In figure 5 an example of a belief network referred to a gas turbine is presented. This network has four nodes expressing the parameters of the engine taken into account. These are: the 'efficiency factor of the high pressure compressor' (n(HPC)), the 'efficiency factor of the high pressure turbine' (n(HPT)), the 'pressure ratio' (π_c) and the 'turbine inlet temperature' (TIT).

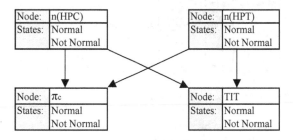

Figure 5. An example of a belief network of a gas turbine.

Each node has two or more discrete states, expressing all the different states of the parameter they refer to. For instance, in the network of figure 5, node TIT has two states: 'normal temperature' and 'not normal temperature'. In each case, the set of states of a node must be exhaustive and mutually exclusive. In other words, any possible condition of a parameter expressed by a node in a BBN is represented by one and only one state of this node. Links among the nodes express the 'rules' of interdependence that hold among them. For example, the link from node n(HPC), on the network of figure 5, to node πc expresses the fact that the condition (state) of node n(HPC) affects directly the condition of node πc. The absence of a link between two nodes doesn't mean that these two nodes are independent, but expresses the fact that the condition (state) of the one doesn't directly affect the condition of the other.

Each node has a Conditional Probability Table (CPT), expressing the probability each state of the node to occur, when the state of each other node, ending up directly to it (called 'parent node'), is known. In case that, a node has no other nodes ending up directly to it (called a 'root node'), the CPT of this node express the 'a priori' probability each state of this node to occur. In Table 1 an example of how the CPTs, of the nodes of the network of figure 5, could be, is shown.

Node:	n(HPC)	
		Not
States:	Normal	Normal
a priori probabilities	0.90	0.10

Node:	n(HPT)	
		Not
States:	Normal	Normal
a priori probabilities	0.87	0.13

Node	TE				Parent
States	Normal	Not Normal	n(HPC)	n(HPT)	Nodes
Conditional Probabilities	0.97	0.03	Normal	Normal	States of parents
	0.68	0.32	Normal	Not Normal	
	0.26	0.74	Not Normal	Normal	
	0.02	0.98	Not Normal	Not Normal	

Node	TIT				Parent
States	Normal	Not Normal	n(HPC)	n(HPT)	Nodes
Conditional Probabilities	0.97	0.03	Normal	Normal	States of parents
	0.16	0.84	Normal	Not Normal	
	0.38	0.62	Not Normal	Normal	
	0.08	0.92	Not Normal	Not Normal	

Table 1. An example of the CPTs of the nodes of the network of figure 5.

Once a BBN is constructed, inference can be realized any time evidence is available. Inference is the procedure where the probabilities of each state of each node of the network are updated each time that evidence is available. 'Evidence' is the knowledge of the state of one or more nodes of the network.

Bayesian Belief Networks have some features that make them very attractive in the field of diagnosis of faults in gas turbines. The most important of these features are: BBN allow probabilistic diagnosis; it is more realistic to make diagnosis expressing the belief (probability) of whether an event occurred or not, than expressing a deterministic answer. Mathematical relationships among the variables of a network are not required in order to form a BBN. Only the way that these variables affect each other is required. This is very helpful since such mathematical relationships may be unknown. Modern approximate algorithms for inference with BBN are able, nowadays, to answer queries, once 'evidence' is provided, within few seconds, even for complicated networks, performing with adequate accuracy. Each node of a BBN can be an 'evidence' as well as a 'query'. There is no restriction to the number of 'query' or 'evidence' nodes. Therefore, there is no limitation on how many or which are the 'evidence' nodes in order to estimate the probabilities of all the other nodes of a network. It allows also the inclusion of information of different nature and from different sources for diagnostics.

Such networks have been employed in the field of gas turbine diagnostics by few researchers. Breese et al. [33], presented a method for detecting specific faults on large gas turbines that combines a thermodynamic model of the engine under examination and a BBN, constructed by use of statistical data of the engine. Palmer [34], presented a statistically also constructed BBN for fault detection of the CF6 family of engines.

The first attempt to propose a general procedure of building a BBN for diagnostic purposes, has been presented by Romessis et al. [35]. The objective of the investigation was to reveal a possible way of setting up such a network with aid of an engine performance model. The way of building diagnostic BBNs, allowing implementation into any type of engine, and the disengagement of the BBN from hard to find statistical data, were two elements that made the work interesting and promising. The effectiveness of the proposed diagnostic method was examined on benchmark fault case scenarios, in a typical modern turbofan engine of civil aviation. The diagnosis was based on the observation of fewer measurements (7) than the considered fault parameters (11). Inference with BBN showed that such a network is very reliable, since in the 96% of the cases where a fault was detected, it was detected correctly. Only a 4% of the cases were attributed to a wrong fault.

A more efficient method even in fault cases with smaller health parameters' deviations was proposed in [36]. The improvement was due to the way the BBN is constructed: probabilistic relationships among variables are more accurately represented. The effectiveness of the proposed method has been demonstrated by its strong diagnostic ability with various fault scenarios and cases at several operating conditions, including coverage of an operational envelope of a typical flight.

5. Hybrid and fusion information techniques

Despite research in various methods for engine fault diagnostics, there is still no method which can effectively address all issues. One way to approach the problem is to try and offset the limitations of one technique with the strength of the other. Hybrid models have attempted to bridge this gap.

An integrated fault diagnostics model for identifying shifts in component performance and sensor faults using Genetic Algorithm and Artificial Neural Network was presented in [37]. The diagnostics model operates in two distinct stages. The first stage uses response surfaces for computing objective functions to increase the exploration potential of the search space while easing the computational burden. The second stage uses concept of a hybrid diagnostics model in which a nested neural network is used with genetic algorithm to form a hybrid diagnostics model. The nested neural network functions as a pre-processor or filter to reduce the number of fault classes to be explored by the genetic algorithm based diagnostics model. The hybrid model improves the accuracy, reliability and consistency of the results obtained. In addition significant improvements in the total run time have also been observed. Ecstase [38], presents an example of the use of fuzzy logic combined with influence coefficients applied to engine test-cell data to diagnose gas-path related performance faults. The diagnostic process to identify module level engine performance faults has been validated using eight examples from real-world test-cell data. Many combinations of faults were examined in an attempt to explain the performance degradation observed in the engine under- going repair. This aspect of the process enabled the status of 17 faults to be determined, despite only five engine parameters being used. The method correctly identified the faults for all except for one fault which had a very small degradation effect on the engine performance.

A diagnostic method consisting of a combination of Kalman filters and Bayesian Belief Network (BBN) is presented in [39]. A soft-constrained Kalman filter uses a priori information derived by a BBN at each time step, to derive estimations of the unknown health parameters. The resulting algorithm has improved identification capability in comparison to the stand-alone Kalman filter. Besides the improvements in accuracy and stability, this kind of method allows information or sensor fusion, which is a very important field of research for future works. The key advantage of combining methods is that it replaces the problem of comparing classification techniques to regression techniques by the problem of choosing which information they can share. Romessis et al. [40] proposed a statistical processing of the diagnostic conclusions provided by a least-square based gas path diagnostic method, in order to improve diagnosis. In a similar attempt (see [41]) a combinatorial approach (statistical evaluation of least squares estimations) combined with fuzzy logic rules to calculate fault probabilities. The possibility of creating a mixed fault classification that incorporates both model-based and data driven fault classes was investigated in [42]. Such a classification combines a common diagnosis with a higher diagnostic accuracy for the data-driven classes. The performed analysis has revealed no limitations for realizing a principle of the mixed classification in real monitoring systems.

Information Fusion is the integration of data or information from multiple sources, to achieve improved accuracy and more specific inferences than can be obtained from the use of a single sensor alone. It is generally believed that an ensemble of methods improves diagnostic accuracy when compared to individual methods. In [43] several fusion architectures and classifiers were evaluated. Fusing classifiers that are performing very well had little positive effect. However, it was shown that fusing marginal classifiers can increase the diagnostic performance substantially, while reducing their variability. Enhanced fault localization using probabilistic fusion with gas path analysis algorithms is referred in [44],while a fusion technique allowing the merge of conclusions provided by diagnostic methods that act independently for the detection of gas turbine faults is described in [45]. The proposed technique adopts the principles of Dempster-Schafer theory for the fusion of two diagnostic methods namely a Bayesian Belief Networks (BBN) and a Probabilistic Neural Networks (PNN). The technique has been applied for the detection of thermodynamic as well as mechanical faults on gas turbines. In all cases, the effectiveness of the proposed fusion technique demonstrated that the merge of diagnostic information from different sources leads to better and safer diagnosis.

A fusion method that utilizes performance data and vibration measurements for gas turbine component fault identification is presented in [46]. The proposed method operates during the diagnostic processing of available data (process level) and adopts the principles of certainty factors theory. Both performance and vibration measurements are analyzed separately, in a first step, and their results are transformed into a common form of probabilities. These forms are interwoven, in order to derive a set of possible faulty components prior to deriving a final diagnostic decision. Then, in the second step, a new diagnostic problem is formulated and a final set of faulty health parameters are defined with higher confidence. In the proposed method the non-linear gas path analysis is the core diagnostic method, while information provided by vibration measurements trends is used to narrow the domain of unknown health parameters and lead to a well-defined solution. Finally a comprehensive presentation of different fusion possibilities offered is given in [47].

6. ECMD integrated systems

Although many diagnostic methods have been proposed and some of them have been tested in real engines only few are known to be incorporated in ECMD integrated systems. An industrial monitoring and diagnostic system must comply with several requirements. For such a system to be effective it should:

- Be as automated as possible and integrated namely performing all actions from data collection to derivation of diagnostic decisions.

- Be "robust", namely not very susceptible to noise or faulty input information.

- Have an as wide as possible coverage of detectable faults. Additionally, it should allow additions of other newly discovered faults, which have not been included in the initial repertory of the system.

- Have prognostic capabilities concerning future maintenance and repair actions. This helps in ensuring that long lead-time spares are available and that outages be minimized.

- Derive information with high confidence. In this respect, derivation of the same conclusion by different methods is a very useful feature.

- Employ as few instruments as possible. The instrumentation should be kept as simple as possible and include the minimum number of instruments.

- Be modular and flexible with open circuit architecture in order to be adapted to operator's needs.

- Be very user friendly, so that it can be used by non-specialized personnel, while its output is clear enough to need very little or no interpretation.

In order to materialize a monitoring system, which possesses these features, the procedures, which should be implemented, are as follows:

i.   Measurement data acquisition.

ii.   Data evaluation in order to discard unreliable readings and possibly detect sensor faults.

iii.   Data processing using appropriate techniques in order to derive diagnostic information.

iv.   Diagnostic inference in order to decide what is the nature, the location and the se-
          verity of a malfunction present, if any.

v.   Data management in order to keep historical data records for long term monitor-
          ing, without storing too much unnecessary information.

Such an integrated system and experience gained from its implementation on an operating
industrial gas turbine has been presented in [48]. The main functions of this system materializing the procedures mentioned above are as follows:

Data Acquisition and Management: Data are acquired from a number of different measuring instruments, for slowly or fast varying quantities. The obtained measurements are being
on-line validated and then organized in a database. The system also gives the possibility to
play back measurements database in order to recreate real time operation. Additional features of the developed data acquisition feature are its flexibility and its capability to easily
meet the requirements of any particular implementation.

Performance Analysis: The acquired thermodynamic measurements are being on-line proc-
essed using the adaptive modeling method [19]). Thus, at any given operating conditions,
the overall engine performances and individual components health indices are being evaluated. The method can also be used off-line for the analysis of previously recorded data.

EGT Monitoring: The hot section, being the most critical area of the engine, is receiving particular attention, through exhaust gas temperature profile monitoring ([49]. This monitoring
provides indication of possible burner malfunctions or thermocouple faults. Off- line analysis of historic data stored in measurements database can also be performed.

Vibration Monitoring: The means of identifying mechanical faults are provided by this
function of the system. For data from vibration sensors, the following diagnostic features are
extracted and assessed: a) overall vibration level, b) power spectra (on-line frequency analysis), and c) spectral signatures [50]). Finally, as the other monitoring modules, it offers the
possibility of off-line line analysis of historic data stored in measurements database.

These functions are performed by the system continuously, while the engine is in operation.
Their implementation provides adequate diagnostic information about engine condition.
This information is being further assessed using a rule based inference engine that provides
an engine condition assessment. Thus, the user is being informed in real time about the engine's condition and performance. The main interface of the system implemented on a PC is
shown in figure 6a. It comprises an axial cut-out of the monitored engine and gives the most
critical information about the engine condition. The system offers the possibility to perform
a more detailed analysis by activating the previously described functions through the but-
tons on the menu at the upper right hand corner.

An example of system effectiveness in diagnosing is the following. A twin-shaft industrial
gas turbine with 21 MW nominal output, used for electricity production in a power station,
is considered. The turbine suffered from the formation of deposits on gas generator and
power turbine blades, very soon after it was put on operation (see figure 1a). A remedy action taken by the manufacturer was a small re-staggering (opening) of power turbine sta-

tionary blades. An easy and reliable way of identification of the malfunction of the turbine is provided by the method of adaptive modeling. The technique has been applied to test data from this turbine and it gave a clear picture of the problem. Comparison of health parameters deviation obtained from data from the initial condition of the engine and after the presence of the problem was detected is shown in figure 6b. It is clearly shown that the swallowing capacity of both turbines has been significantly reduced, as factor f3 shows a reduction of more than 1.5% and f5 more than 3%. The reduction in f1 (of ~ 0.8%) indicates that the compressor has also suffered some deterioration.

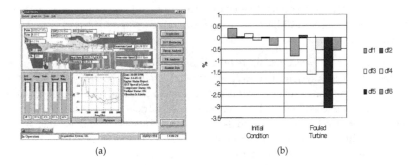

(a)   (b)

Figure 6. (a) Display of a user friendly monitoring software for an industrial gas turbine. (b) Health Indices Percentage deviation, for a gas turbine, which has suffered severe turbine fouling, caused by fuel additives.

7. Conclusions

In this chapter, we have attempted to present basic principles of the engine condition monitoring and diagnostics (ECMD) subject. It would be impossible to cover in few pages all the aspects of ECMD. Thousands of papers have been published and a vast amount of knowledge has been accumulated. Even extensive reviews cannot mention all the proposed methods. In this respect we presented selective methods representative of three main steps of an ECMD approach, namely data acquisition, data processing and diagnostic decision-making, with emphasis on the last two steps. Few recently developed hybrid, data and method fusion techniques have also been briefly discussed. The structure of an integrated ECMD system incorporating different diagnostic technics and already in operation is also presented.

The following conclusions are the outcome of over twenty five years of experience in the area of ECMD.

• The main problems with respect to the industry adoption of advanced technics are the following: a) lack of data due to no data collection and/or data storage at all; b) lack of efficient communication between method developers and maintenance practitioners; c) lack of efficient validation approaches.

- Both physics based and data-driven models show benefits and drawbacks. From the decision making point of view both traditional and artificial intelligence techniques are used, although it seems that hybrid approaches are more promising.

- The value of vibration monitoring and other sources data in refining gas-path monitoring results has been recognized. The approach of combining different monitoring results, i.e., data fusion, is becoming an active area of research.

- Usage and life monitoring for fatigue critical or life-limited parts are increasingly important

- Collaboration of ECMD research groups is necessary in order to produce integrated platforms for enhancing an ECMD system since each research group has its own specialty and focus in the area.

The following research directions are required for the next generation of ECMD systems: Enhancement of ECMD systems to collect accurate information, especially fault event information. This information would be very useful for model building and model validation as well. Advanced models and methods for utilization of the transient data diagnostic information as well as detailed higher order models for deterioration mechanisms and faults reliable simulation should be developed. Accurate prognostic models development is also necessary. Finally, there is a need for establishment of efficient validation approaches through benchmark test cases to compare the merits and the drawbacks of different modeling and algorithmic approaches.

Author details

Anastassios G. Stamatis

Address all correspondence to: tastamat@uth.gr

Mechanical Engineering Department, Polytechnic School, University of Thessaly, Volos, Greece

References

[1] Zaita AV, Buley G, Karlsons G. Performance deterioration modeling in aircraft gas turbine engines. Journal of Engineering for Gas Turbines and Power. 1998;120(2): 344-9.

[2] Kurz R, Brun K. Degradation in gas turbine systems. Journal of Engineering for Gas Turbines and Power. 2001;123(1):70-7.

[3] Diakunchak IS. Performance Deterioration in Industrial Gas-Turbines. Journal of Engineering for Gas Turbines and Power. 1992;114(2):161-8.

[4] Li YG. Performance-analysis-based gas turbine diagnostics: A review. Proceedings of the Institution of Mechanical Engineers, Part A: Journal of Power and Energy. 2002;216(5):363-77.

[5] Marinai L, Probert D, Singh R. Prospects for aero gas-turbine diagnostics: A review. Applied Energy. 2004;79(1):109-26.

[6] Urban LA, Volponi AJ. Mathematical methods of relative engine performance diagnostics. SAE 1992 transactions, vol. 101, journal of aerospace, technical paper 922048.

[7] Barwell MJ. Compass - Ground Based Engine Monitoring Program for General Application. 1987. SAE technical paper 871734.

[8] Doel DL. TEMPER - a gas-path analysis tool for commercial jet engines. Journal of Engineering for Gas Turbines and Power. 1994;116(1):82-9.

[9] Doel DL. An assessment of weighted-least-squares-based gas path analysis. J Eng Gas Turbines Power, Trans ASME. 1994;116:365-73.

[10] Stamatis A, Papailiou KD. Discrete operating condition gas path analysis. AGARD-CP-448, Engine Condition Monitoring - Technology and Experience. 1988.

[11] Gronstedt TV. A multi point gas path analysis tool for gas turbine engines with a moderate level of instrumentation. 2001.XV ISABE, Bangalore, India, Sept. 3-7.

[12] Gulati A, Taylor D, Singh R. Multiple operating point analysis using genetic algorithm optimization for gas turbine diagnostics. 2001. XV ISABE, Bangalore, India, Sept. 3-7

[13] Pinelli M, Spina PR, Venturini M. Optimized Operating Point Selection for Gas Turbine Health State Analysis by Using a Multi-Point Technique. ASME Turbo Expo 2003, collocated with the 2003 International Joint Power Generation Conference (GT2003) June 16–19, 2003, Atlanta, Georgia, USA, Paper no. GT2003-38191.

[14] Mathioudakis K, Kamboukos P. Assessment of the effectiveness of gas path diagnostic schemes. Journal of Engineering for Gas Turbines and Power. 2006;128(1):57-63.

[15] Gronstedt T. Identifiability in multi-point gas turbine parameter estimation problems. ASME Turbo Expo 2002: Power for Land, Sea, and Air (GT2002) June 3–6, 2002, Amsterdam, The Netherlands, Paper no. GT2002-30020.

[16] Henriksson M, Borguet S, Leonard O, Gronstedt T. On inverse problems in turbine engine parameter estimation.. ASME Paper no. GT2007-27756

[17] Aker GF, Saravanamuttoo HIH. Predicting gas turbine degradation due to compressor fouling using computer simulation techniques. ASME paper 88-GT-206.

[18] Stamatis AG. Evaluation of gas path analysis methods for gas turbine diagnosis. Journal of Mechanical Science and Technology. 2011;25(2):469-77.

[19] Stamatis A, Mathioudakis K, Papailiou KD. Adaptive simulation of gas turbine performance. Journal of Engineering for Gas Turbines and Power. 1990;112(2):168-75.

[20] Li YG, Korakianitis T. Nonlinear weighted-least-squares estimation approach for gas-turbine diagnostic applications. Journal of Propulsion and Power. 2011;27(2): 337-45.

[21] Ogaji S, Sampath S, Singh R, Probert D. Novel approach for improving power-plant availability using advanced engine diagnostics. Applied Energy. 2002;72(1):389-407.

[22] Zedda M, Singh R. Gas turbine engine and sensor fault diagnosis using optimization techniques. Journal of Propulsion and Power. 2002;18(5):1019-25.

[23] Mathioudakis K, Kamboukos P, Stamatis A. Gas turbine component fault detection from a limited number of measurements. Proceedings of the Institution of Mechanical Engineers, Part A: Journal of Power and Energy. 2004;218(8):609-18.

[24] Grodent M, Navez A. Engine Physical Diagnosis Using a Robust Parameter Estimation Method., 37th AIAA/ASME /SAE/ASEE Joint Propulsion Conference and Exhibit, 8-11 July 2001, Salt Lake City, Utah, paper AIAA-2001-3768.

[25] Mathioudakis K, Kamboukos P, Stamatis A. Turbofan performance deterioration tracking using nonlinear models and optimization techniques. Journal of Turbomachinery. 2002;124(4):580-7.

[26] Stamatis A, Mathioudakis K, Papailiou K. Optimal measurement and health index selection for gas turbine performance status and fault diagnosis. Journal of Engineering for Gas Turbines and Power. 1992;114(2):209-16.

[27] Ogaji SOT, Sampath S, Singh R, Probert SD. Parameter selection for diagnosing a gas-turbine's performance-deterioration. Applied Energy. 2002;73(1):25-46.

[28] Eustace R, Merrington G. Fault diagnosis of fleet engines using neural networks. Fault Diagnosis of Fleet Engines Using Neural Networks. ISABE paper, ISABE 95-7085.

[29] Volponi AJ, DePold H, Ganguli R, Daguang C. The use of kalman filter and neural network methodologies in gas turbine performance diagnostics: A comparative study. Journal of Engineering for Gas Turbines and Power. 2003;125(4):917-24.

[30] Kanelopoulos K, Stamatis A, Mathioudakis K. Incorporating neural networks into gas turbine performance diagnostics. ASME paper 97-GT-35.

[31] Romessis C, Stamatis A, Mathioudakis K. A parametric investigation of the diagnostic ability of probabilistic neural networks on turbofan engines. ASME paper 2001-GT-0011.

[32] Romesis C, Mathioudakis K. Setting up of a probabilistic neural network for sensor fault detection including operation with component faults. Journal of Engineering for Gas Turbines and Power. 2003;125(3):634-41.

[33] Breese JS, Gay R, Quentin GH. Automated decision-analytic diagnosis of thermal performance in gas turbines. ASME paper 92-GT-399.

[34] Palmer CA. Combining Bayesian belief networks with gas path analysis for test cell diagnostics and overhaul. ASME paper 98-GT-168.

[35] Romessis A, Stamatis A, Mathioudakis K. Setting up a belief network for turbofan diagnosis with the aid of an engine performance model. ISABE-2001-1032.

[36] Romessis C, Mathioudakis K. Bayesian network approach for gas path fault diagnosis. Journal of Engineering for Gas Turbines and Power. 2006;128(1):64-72.

[37] Sampath S, Singh R. An integrated fault diagnostics model using genetic algorithm and neural networks. Journal of Engineering for Gas Turbines and Power,2006;128, (1):49-56.

[38] Eustace RW. A real-world application of fuzzy logic and influence coefficients for gas turbine performance diagnostics. Journal of Engineering for Gas Turbines and Power. 2008;130(6).

[39] Dewallef P, Romessis C, Leonard O, Mathioudakis K. Combining classification techniques with Kalman filters for aircraft engine diagnostics. Journal of Engineering for Gas Turbines and Power. 2006;128(2):281-7.

[40] Romessis C, Kamboukos P, Mathioudakis K. The use of probabilistic reasoning to improve least squares based gas path diagnostics. Journal of Engineering for Gas Turbines and Power. 2007;129(4):970-6.

[41] Lipowsky H, Staudacher S, Nagy D, Bauer M. Gas turbine fault diagnostics using a fusion of least squares estimations and fuzzy logic rules. ASME Turbo Expo 2008: Power for Land, Sea, and Air (GT2008)June 9–13, 2008, Berlin, Germany. ASME Paper no. GT2008-50190

[42] Loboda I, Yepifanov S. A mixed data-driven and model based fault classification for gas turbine diagnosis. Proceedings of ASME Turbo Expo 2010: International Technical Congress, 8p., Scotland, UK, June 14-18, Glasgow, ASME Paper No. GT2010-23075.

[43] Donat W, Choi K, An W, Singh S, Pattipati K. Data visualization, data reduction and classifier fusion for intelligent fault diagnosis in gas turbine engines. Journal of Engineering for Gas Turbines and Power. 2008;130(4).

[44] Kyriazis A, Mathioudakis K. Enhanced fault localization using probabilistic fusion with gas path analysis algorithms. Journal of Engineering for Gas Turbines and Power. 2009;131(5).

[45] Romessis C, Kyriazis A, Mathioudakis K. Fusion of gas turbines diagnostic inference - The dempster-schafer approach. Proceedings of IGTI/ASME Turbo Expo 2007, 9p., Canada, May 14-17, 2007, Montreal, ASME Paper GT2007-27043.

[46] Kyriazis A, Tsalavoutas A, Mathioudakis K, Bauer M, Johanssen O. Gas turbine fault identification by fusing vibration trending and gas path analysis. ASME Turbo Expo

2009: Power for Land, Sea, and Air (GT2009),June 8–12, 2009, Orlando, Florida, USA, ASME Paper no. GT2009-59942

[47] Volponi AJ, Brotherton T, Luppold R. Development of an information fusion system for engine diagnostics and health management. AIAA 1st Intelligent Systems Technical Conference20 - 22 September 2004, Chicago, Illinois, AIAA Paper 2004-6461.

[48] Mathioudakis K, Stamatis A, Tsalavoutas A, Aretakis N. Performance analysis of industrial gas turbines for engine condition monitoring. Proceedings of the Institution of Mechanical Engineers, Part A: Journal of Power and Energy. 2001;215(2):173-84.

[49] Tsalavoutas A, Mathioudakis K, Smith MK. Processing of circumferential temperature distributions for the detection of gas turbine burner malfunctions. ASME Paper 96-GT-103.

[50] Loukis E, Wetta P, Mathioudakis K, Papathanasiou A, Papailiou A. Combination of different unsteady quantity measurements for gas turbine blade fault diagnosis. ASME paper, 91-GT-201, International Gas Turbine and Aeroengine Congress and Exposition, June 3-6 1991, Orlando.

Chapter 10

Thermal Shock and Cycling Behavior of Thermal Barrier Coatings (TBCs) Used in Gas Turbines

Abdullah Cahit Karaoglanli, Kazuhiro Ogawa,
Ahmet Türk and Ismail Ozdemir

Additional information is available at the end of the chapter

1. Introduction

Gas turbine engines work as a power generating facility and are used in aviation industry to provide thrust by converting combustion products into kinetic energy [1-3]. Basic concerns regarding the improvements in modern gas turbine engines are higher efficiency and performance. Increase in power and efficiency of gas turbine engines can be achieved through increase in turbine inlet temperatures [1,4]. For this purpose, the materials used should have perfect mechanical strength and corrosion resistance and thus be able to work under aggressive environments and high temperatures [2]. The temperatures that turbine blades are exposed to can be close to the melting point of the superalloys. For this reason, internal cooling by cooling channels and insulation by thermal barrier coatings (TBCs) is used in order to lower the temperature of turbine blades and prevent the failure of superalloy substrates [1-4]. By utilizing TBCs in gas turbines, higher turbine inlet temperatures are allowed and as a result an increase in turbine efficiency is obtained [5]. TBCs are employed in a variety of areas such as power plants, advanced turbo engine combustion chambers, turbine blades, vanes and are often used under high thermal loads [6-11]. Various thermal shock tests are conducted by aerospace and land gas turbine manufacturers in order to develop TBCs and investigate the quality control characteristics. Despite that fact, a standardized method is still lacking. The reason lies behind the difficulty of finding a testing method that can simulate all the service and loading conditions. Present testing systems developed by the engine manufacturers for simulation of real thermal conditions in engines consist of; burner rig thermal shock testing units, jet engine thermal shock testing units and furnace cycle tests [16-20]. In this study, thermal cycle and thermal shock behavior of TBC systems under service conditions are examined, and a collection of testing methods used in evaluation of performance and endurance properties and recent studies regarding aforementioned concerns are presented as a review Study consists of the following chapters; 1. Introduction, 2. Thermal Barrier Coatings (TBCs), 2.1 An Overview of TBCs, 2.2 Structure and function of TBC systems, 2.2.1 Substrate material, 2.2.2 Bond Coat, 2.2.3 Top Coat, 3. Thermal Shock and Cycling Behavior of Thermal Barrier Coatings, 3.1 Thermal Shock Concept, 3.2 Thermal Cycle/Shock Tests for TBCs, 4. Summary, 5. Acknowledgment, 6. References.

Including the introductory chapter, the study consists of four parts;

1.Chapter: The aim of the study is explained. An introduction is given as; general characteristics and application of thermal barrier coatings in gas turbine engines, and thermal cycle/shock characteristics under service conditions.

2.Chapter: Thermal Barrier Coating (TBC) systems are presented and also production, structure and characteristics are explained.

3.Chapter: Thermal shock and cycle behavior of TBC system applications in gas turbines is given. Testing methods and criteria is presented. Evaluation of TBC systems after thermal shock/cycle tests is given and microstructural evaluation is mentioned.

4.Chapter: The findings of given studies are summarized and results are presented.

2. Thermal barrier coating (TBC)

2.1. An overview of TBCs

A typical TBC system, which is used in gas turbine engines to thermally protect metallic components from aggressive environmental effects, consists of a superalloy substrate material, a metallic bond coat for oxidation resistance, a ceramic top coat (such as ZrO_2 stabilised with % 6-8 Y_2O_3) for thermal insulation and a thermally grown oxide layer (TGO) that forms at the bond coat-top coat interface as a result of bond coat oxidation in service conditions [2,15,21].

2.2. Structure and function of TBC systems

The main function of TBCs is to provide thermal insulation against hot gasses in engines and turbines and thus reduce the surface temperature of the underlying alloy components [21-22]. To do this, while the coated parts are cooled inside, the heat transfer through TBC to the component should be kept low. With approximately 300 μm thick YSZ top coat, it is possible to achieve a temperature drop up to 170 °C between the top coat surface and substrate [22-24]. Figure 1 shows a TBC system applied on the turbine vane and its temperature gradient.

Heat insulation property of TBCs can be utilised in gas turbines in two different ways. In turbines where a TBC system is applied, either the service life of the component is increased by keeping the working temperature of the engine unchanged and thus decreasing temperature of the underlying substrate, or the efficiency is increased by increasing the working temperature of the engine to a level at which the temperature of the coated substrate is same as the uncoated substrate temperature.[23]. TBC systems that are produced in two different ways with conventional methods are shown in Figure 2 [25].

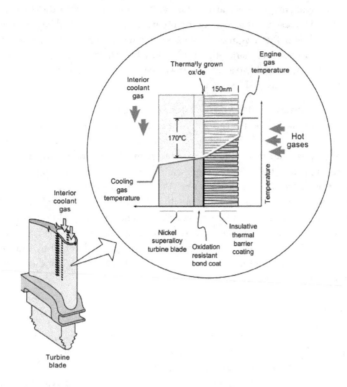

Figure 1. Representation of a TBC structure which is applied to turbine vane to serve as a thermal insulator and the heat gradient in the system [24].

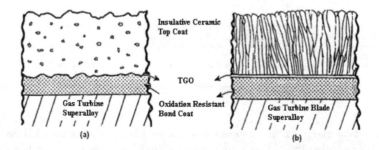

Figure 2. TBC structures produced with different methods: a) produced by APS method, b) produced by EB-PVD method [25].

Top layer is employed to achieve the desired temperature reduction. The lower the heat that crosses the ceramic top layer is, the more effective the cooling and hence the lower the component's surface temperature will be. To achieve this goal, the top layer should be chosen from a material with a low thermal conductivity. Another way to decrease the thermal conductivity is to increase the thickness of top layer. However it should be considered that by an increase in thickness, the weight of the component and the residual stresses in the coating will also increase. In addition, since the heat conduction distance is higher in a thicker top layer, heat transfer rate will decrease which may result in a surface temperature that exceeds the ceramic materials limits [22].

The temperature decrease with the use of TBCs provides many advantages. First of all, with the decrease in the rate of the heat transferred to the component, service temperature and indirectly productivity can be increased. Or by decreasing the temperature on the component, the substrate material that forms the component is enabled to show properties close to the room temperature properties. Besides, creep can be reduced with the component's temperature decrease as well. In addition, by means of TBCs, the protection against chemical damages, such as oxidation, is achieved by reducing the oxidation rate through the reduction in temperature and appropriate bond coat material selection [26-28]. How TBCs perform the mentioned tasks can be better understood by examination of the materials and structures of the layers that form TBC. General structure of TBCs is explained below by examining every layer (i.e. substrate, bond and top coat layers and TGO that forms by bond coat oxidation) in detail and according to their functions.

2.2.1. Substrate material

Substrate is in fact the basic material already available in coating system and the coating is placed on it. So, substrate is the main element to be intended to protect. Ni based superalloys are generally used in gas turbines as substrate material. The main reason for this selection is that superalloys can protect their strength under high temperatures such as 2000 °F (~1100 °C). In order to increase the creep resistance at high temperatures, substrate is produced with directional grains or single crystal structure [22, 29-30]. A general composition of a conventionally used Inconel 718 super alloy is given in Table 1 [31].

% Chemical Composition								
Cr	Ni	Nb	Mo	Ti	Al	Cu	C	Fe
19.0	52.5	5.1	3.0	0.9	0.5	0.15 max.	0.08 max.	Balance

Table 1. Chemical composition of Inconel 718 superalloy [31].

While the working temperatures of superalloys are quite high, coatings are used in today's gas turbines to increase working temperature in the turbine even higher and to extend the service life of the parts/components. As can be seen in Figure 3, working temperatures of gas turbines are already so close to the melting temperature of elements comprising superalloy components [32].

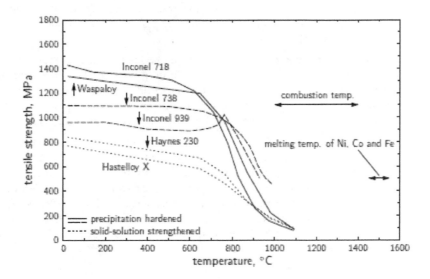

Figure 3. Tensile strength of some superalloys as a function of temperature [32].

Because of the various environmental conditions that turbine blades are exposed to, turbine inlet temperatures have greatly increased since 1940s. Today's commercial and military aircrafts have turbine inlet temperature respectively over 1500°C and 1600°C and are expected to reach 1760 °C or more at the end of 2015, and this obviously shows the need for thermal barrier coatings. Turbine blades work under much harder conditions than any other component in the engine due to high temperature and stresses they are exposed to and the rapid temperature changes they undergo during (thermal) cycles. Moreover, they are also faced with oxidation and corrosion due to hot gasses and chemicals in the working environment. Because of all of these reasons, turbine blade components should have properties such as high corrosion resistance, creep resistance, and fatigue strength in the service. In order to meet these properties, a large proportion of the materials used in making of today's modern airplane gas turbine engines consist of superalloy materials [31].

2.2.2. Bond coat

Bond coat has two main functions in TBC systems. First of these functions is to increase the adherence between ceramic top coat and substrate. Second function, which cannot be performed by top coat due to its porous structure, is to protect the underlying material from chemical attacks such as oxidation [26,33]. In order for the bond coat to continue its first function, a material with suitable thermal expansion ratio should be selected [24]. This way, stresses which occur between top coat and substrate because of the thermal expansion and shrinkage during heating and cooling, can be kept at a minimum. Considering that bond coats are conventionally produced from metal alloys with high thermal expansion coefficients and that top coats are produced from ceramics with low thermal expansion coefficient, the tension between these surfaces should be expected to decrease by a decrease in expansion coefficient of the bond coat material [34].

Porous structure of the top coat and high diffusivity of oxygen ion in this layer enables the surface oxygen to reach lower layers [35]. Thus, it is the duty of the bond coat to protect the substrate against chemical attacks like oxidation. In order to fulfil this duty, bond coat contacts with oxygen and creates an oxide layer on top coat and interface surface. This layer, which is thinner than 10 μm and forms on the bond coat surface during service, is called TGO [23].

Considering the mechanisms mentioned in this part, TGO layer is desired to consist of a homogeneously distributed, continuous and dense α-Al_2O_3 [36]. However, there will be various spinel and metallic oxides apart from alumina in such a structure. In fact, oxides other than α-Al_2O_3 are seen to form in time at TGO layer [37]. The reason why TGO is desired to consist of α-Al_2O_3 is that oxygen permeability of this alumina phase is low [36,38]. Because if an oxide layer has low oxygen permeability, growth rate will also be low and failure stemming from TGO will be postponed. Material selection in bond coat should be designed suitably in order to achieve the above-mentioned properties.

2.2.3. Top coat

Top coat is the outermost layer, which contacts with the hot working gasses in gas turbine and so is exposed to the engine's working temperature. The basic function of top coat is to provide thermal isolation to the underlying layers [31,39]. A top coat should have some basic properties to achieve this objective. These properties are; high melting temperature (to keep coating structure when in contact with hot gasses), low thermal conductivity (to fulfil its thermal insulation function), thermal expansion coefficient in accordance with the underlying superalloy (to prevent the mismatch between layers during thermal cycles), resistance to oxidation and corrosion (be-cause service environment include oxygen and some other gasses at high temperature), strain tolerance (in order to resist thermal shocks during thermal cycles) [22,40-41].

Most of the properties above are general characteristic properties of ceramics. A ceramic material that includes third and fifth properties as well will be a suitable material for top coat. Conventionally, top coat consists of a tetragonal structured zirconia. Pure zirconia undergoes phase transformation at 1170 °C and forms a monoclinic phase by diffusionless transformation. This situation causes a volum expansion of about %4 [42]. Volume change is undesirable because it may cause tensile stresses in the material. Therefore, to avoid transformation from tetragonal phase to monoclinic phase, yttria is added to zirconia. By doing so, metastable tetragonal phase of zirconia is formed and tetragonal phase is stabilised in low temperatures. This metastable tetragonal phase will not transform to monoclinic phase in low temperatures. But if sufficient time and temperature is provided, it transforms to stable tetragonal phase and cubic phase. Stable tetragonal phase that forms under this condition can than transform to monoclinic phase under low temperatures [43-44]. The basic property that makes YSZ a suitable material for top coat is that, along with its high thermal stability, it has low thermal conductivity and high thermal expansion coefficient. Unlike ceramics like Al_2O_3 that are unstable at high temperatures due to their polymorph properties, YSZ material has a highly stable structure. [45-46]. As shown in the Figure 4, while YSZ's thermal conductivity is low with respect to ceramics such as Al_2O_3 and MgO, its thermal expansion ratio is higher than ceramics such as SiO_2 or mullite that has low thermal conductivity[24,45].

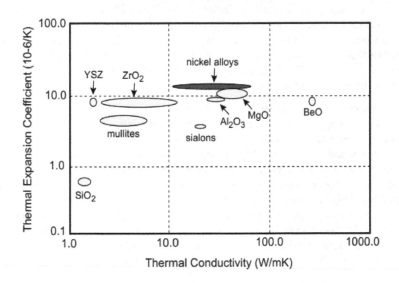

Figure 4. Representation of thermal expansion coefficient and thermal conductivity properties of various materials [24].

According to Figure 4, while thermal expansion ratio of Ni based substrate alloys is $14\text{-}16*10^{-6}$ K^{-1}, thermal expansion ratio of YSZ is about $9*10^{-6}$ K^{-1}. Considering that working temperatures in gas turbines can be as high as 1400°C and it undergoes thermal cycle during service period, it can easily be understood how thermal expansion mismatch can cause failure and how important it is for the top coat expansion coefficient to be close to bond coat [27,32]. With these properties, top coat can provide only the first of the two basic functions of TBC systems, which is heat insulation. Besides, the protection of top coat against corrosion and oxidation remain as an issue due to high oxygen permeability of this layer. The main reason of high oxygen permeability in zirconia top coat is high gas permeability due to microcracks and porosities. However, ionic diffusion can also contribute to oxygen permeability [35,47]. When the high working temperatures of the engine are taken into account, the chemical damages that are caused by the penetrating gases may reach significant levels. Differences in strain tolerances may occur according to deposition method. While tolerance in plasma spray coatings is related to porosities between splats and voids like cracks, tolerance in coatings produced by EB-PVD is related to columnar growth and unattached columns [12,48]. When all these are taken into account, YSZ materials can be seen to be suitable for production of the top coat of TBCs for gas turbine components and superalloy parts with both APS and EB-PVD methods [49].

3. Thermal cycle/shock behaviour of thermal barrier coatings

Performance of TBC systems are closely related to the methods used in production. Plasma spray and EB-PVD methods are widely used in top coat production in TBCs and applied to gas turbine blade and vanes in aviation industry. The service life and mechanic properties of TBCs are closely related to the ceramic top coat microstructure. Characteristic properties in microstructure that stem from coating method in plasma spray coatings have direct effect on thermal cycle/shock behaviour and performance of TBC systems. It is known that microstructure of coatings produced by plasma spray method consist of splats and there are pores, cracks and spaces between lamellas [12-15].

Porosity percentage of ceramic coatings produced with plasma spray method range from %3 to %20. High porosity is an advantage since it reduces the thermal conductivity of the coating. Residual stresses, which occur in YSZ coatings, stem from the thermal expansion mismatch between metal and ceramic. As the porosity in coating increases, the residual stress will decrease [50-52]. Another factor that is effective in coating performance is micro crack density. Micro cracks form as a result of rapid cooling of melted splats in plasma spray ceramic coatings. As the density of horizontal cracks on coating increases, thermal cycle/shock life of coating decreases. As a result, properties such as; porosity, horizontal and vertical cracks and elastic modulus in TBC systems are key parameters that affect thermal cycling life. It is important to keep these parameters in optimum levels in service and to identify their relationship with each other carefully, for the coating system to resist thermal cycling [53-55].

Macro cracks that form perpendicular to the substrate surface of TBC are called segmentation cracks. Coating structures with segmentation cracks have superior properties to other coating structures. Segmentation cracks are known to increase tolerance to stresses that arise from thermal expansion mismatch between substrate and coating. Segmentation cracks increase the stress tolerance of the coating and as a result, significantly decreases thermo-mechanical property differences that cause thermal stresses at substrate and coating interface. Therefore, TBC systems with segmentation cracks show a promising potential for increasing thermal cycling performance and life [56-58].

Ceramic top coats that are applied to aviation components such as turbine blade and vanes and jet engine parts by APS technique need to have high thermal cycle/shock resistance in order to stand high loading conditions. APS coatings mostly fail by spallation due to stress energy that occurs during thermal cycle process. One way to decrease the accumulation of stress is to use coatings with high porosity, because micro-cracks and porosity on coatings can absorb some of the stress. Understanding the failure mechanisms that are activated during thermal cycle/shock tests in APS coatings is only possible by investigation of stress levels. At high temperatures, tensile stresses occur as a result of thermal expansion coefficient differences and temporary temperature gradients during rapid thermal cycling between substrate and ceramic layer in APS coatings. Stress relaxation will take place during isothermal hot period and this creates compressive stress at the end of cooling from service temperature to room temperature. The increase in compressive stress will be the main reason of the increase in cracks by causing short cycle life in coatings. Besides, low shrinking stress levels before cooling will cause low compressive stress and thus driving force necessary for the cracks to propagate will be decreased [58-61].

EB-PVD process is a coating method, which is used to apply TBCs to gas turbine engine parts by melting the material that will be coated, evaporating under vacuum and collecting on the substrate material [15,50-52,62]. Coatings that are produced with EB-PVD method have high strain tolerance and their outer surface and TBC-BC interface are quite smooth. Since EB-PVD coatings have high strain tolerance and ability to work under high temperature oxidation conditions, their endurance under flight working conditions is quite high [14,24,63]. EB-PVD coating's columnar microstructure provides remarkable resistance against thermal shocks and mechanical. This enables turbine blades to be used at high pressure and temperatures. Plasma spray coatings show laminar microstructure. This situation causes cracks to form parallel to surface, which affect working life of TBCs. Coatings produced with plasma spray have 0.8-1.0 W/mK thermal conductivity in room temperature. These values are much lower compared to EB-PVD coatings, thermal conductivity of which is 1.5-1.9 W/mK. That means APS coatings provide much better thermal insulation during service [22,64-66]. In recent years, researchers have shown great interest on above-mentioned properties of TBCs in thermal cycling in relation to prolonging service life and endurance [37,51,67-70].

3.1. Thermal shock concept

One of the weakest points of brittle materials like ceramics is that their thermal shock resistance is low. Thermal shock resistance changes with fracture toughness, elastic modulus, poisson's ratio, thermal expansion coefficient and thermal conductivity. Regarding these parameters, stresses that occur due to the temperature difference between centre and surface of a specimen cooled with water or heated rapidly can be found. This situation, where stresses occur under thermal shock conditions and changes that take place during thermal shock are given in Figure 5. Here, ΔT states temperature difference, T_p states temperature at specimen surface and T_z states the temperature at the centre of the specimen[17].

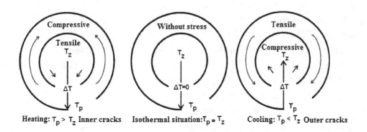

Figure 5. The representation of stress development under different thermal conditions [17].

Ceramic materials, due to their high melting temperature find use in many high temperature applications. In order for the ceramic materials used in TBC systems to resist thermal shock failure, they need to have some basic properties such as; toughness, low thermal conductivity, phase stability at high temperatures, high thermal expansion coefficient and low elastic modulus value [71-72].

Reliability is quite important for TBCs under service conditions. However, since they work under significant temperature fluctuations, some changes in material properties are seen. For example, under normal conditions, gas turbines run and stop repeatedly. This situation brings along degradation mechanisms such as; the thermal expansion, sintering effect and high temperature friction, and thus causes continuous change of the interior stresses in turbine blades. Accordingly, with closing or the growth of cracks, the elastic modulus value changes and this has a major impact on life of TBC under service conditions [55,73-74]. TBC systems can be used as thermal insulators due to their low thermal conductivity. Thermal stresses occur because hot section components that have TBC coating in gas turbines work under rapid thermal cycling conditions in service conditions and this makes the studies rather difficult. Because of this, thermal shock resistance plays an important role in protecting endurance under service conditions in TBCs [75-76].

TBCs fail as a result of removal or separation of coatings under high cycle conditions they are exposed to.

It is believed that the removal of ceramic components under service conditions in TBCs are affected by stresses during service as well as corrosive and erosive degradation damages and residual stress caused by coating process. The increase in thermal shock resistance of coatings that are exposed to thermal cycling can be achieved by controlling residual stresses that occur in service and increasing strain tolerance of ceramic structure. A good resistance can be achieved during thermal cycling by controlling the structural and segmentation micro cracks, and the porosity content [23,61,77].

TBC systems are damaged because of various reasons but failures generally occur as a result of a combination of mechanisms. The failures can take place either in the production of TBC or can take place during service conditions. The basic failure mechanisms that limits the life of TBCs are affected by thermo-mechanic failures, chemical failures, erosion failures, oxidation of bond coating, sintering of top coat, hot erosion effect, CMAS ($CaO-MgO-Al_2O_3-SiO_2$) attack and many other failure types. The most dominant failures mechanism seen in TBCs stems from the formation of TGO structure. A combination of these mechanism with inconsistency in thermal expansion, changes in thermal conductivity ratio and chemical interactions in the engine speed up the failure of TBCs [67,78-82]. Crack formation takes place evantually depending on the time of exposure to high temperature in thermal cycle/shock test. The most important elements that cause the formation of these cracks on TBC and TGO layer are stresses that occur as a result of TGO growth, phase transformations in bond coat, changes in bond coat during thermal cycle and sintering of TBC. Once the cracks form, they propagate and coalesce and result in failure of the coating [83-85].

The formation mechanism of thermo-mechanical stresses change depending on the thermal conditions that TBC is exposed to. If the thermal conditions are isothermal, the mechanism is generally about TGO's growth. But if the TBC is exposed to thermal cycle, the mechanism will be rather related to shrinkage of TGO during cooling. These two situations can be effective in the formation of thermo-mechanic stresses but it should not be ignored that one can dominate the other in some cases. For example; TBCs that work at high temperatures and for long service times are used in gas turbines for energy production on the ground. In this case, isothermal mechanisms become effective and expansion and shrinkage occur when the turbine stops. Consequently, low number of thermal cycle and longer isothermal heating take place in this type of turbines and as a result of this, failure occurs when TGO reaches approximately 5-15 μm thickness. In turbine parts, failures due to thermal expansion mismatch induced by TGO layer and failures due to TGO layer growth are dominant. However, in the turbines that are used in aviation sector where the thermal cycling number is important, isothermal heating is not dominant and failure occur due to thermal cycles when TGO is almost 1-5 μm during service [12,86].

Thermal expansion coefficient mismatch between substrate material and TBC has an important role on the thermal cycle/shock life of TBCs. The rate of mismatch between superalloy substrate material and top coat affects elastic strain energy that is stored during cooling from working temperature. High amount of strain energy causes early removal/breaking of coating as a result of cycling [84,87-88].

Superalloy substrate materials used in TBCs have an effect on thermal cycle life of TBC system. The elements can diffuse from superalloy to bond coat and this diffusion between substrate and bond coat increase or decrease the life depending on the element. For example, as a result of hafnium element diffusion from substrate to bond coat, the adherence of TGO is increased and thus TBC life increases. As a contrary case, the diffusion of tantalum element to bond coat affects the TGO composition and oxides other than alumina may form in TGO structure which results in a reduction of TBC life [84,89-90].

The rapid heating and cooling of coating during thermal cycle inevitably increase the damage on oxide layer. Coating endurance against thermal cycle/shock and degradation can change depending on adherence of coating layers and oxide layer that occurs on coating surface [91-92]. There are three basic reasons of oxide-based removal of coating after thermal shock [92-93]. The first of these reasons reported in the literature is the stress that occurs based on the growth of oxide layer depending on the exposure of the specimen to high temperatures for a long time and removal/breaking and spallation that happen as a result of this. Another factor is the thermal expansion that occur because of the temperature gradient on oxide layer which is a result of rapid heating and cooling. The last factor is the thermal expansion coefficient difference between oxide and coating that take place with the growth of oxide layer. At the end of rapid cooling, compressive stress occur on oxide layer, which has a lower thermal expansion coefficient than substrate material. Stress case changes in rapid heating and tensile stress arise on oxide layer.

Deformations may take place because of the rapid cooling from high temperatures and tensile stress that is generated at the coating/oxide interface [12,92-93].

3.2. Thermal cycle/shock tests for TBCs

For development of TBCs and evaluating the quality of the coatings, the aviation and industrial gas turbine manufacturers apply various thermal cycle/shock tests. TBCs are used usually under high thermal loads in gas turbine parts such as; turbine blades and vanes. There has been no identified method that would provide advantage in comparing the results in this subject. The reason of this is the difficulty of finding a test method that can completely reflect the working conditions. Today, systems that are developed by the engine producers to simulate the real thermal conditions in the engines are burner heating thermal shock test unit (burner rig system), jet engine thermal shock test unit (JETS) and furnace cycle tests. By creating high temperature gradients in ceramics with burner thermal shock test, stresses that affect the integrity of ceramic coating are introduced. Generally disk shaped specimens are used in this test system. The test system is based on cooling of the specimen after heating by a flame where propane and oxygen gases are used together. Since burner heating thermal shock test unit is an expensive system, JETS test has been developed as an economic alternative method for the gradient tests. In JETS test burner equipment is used to create a wide thermal gradient along the TBC and thermo-mechanic stresses on the surface. In furnace cycle oxidation test (FCT) method that is used widely in aviation applications, stresses occur mostly as a result of TGO growth and on ceramic/bond coat.[16-20,68]. A depiction of heating and cooling cycles in burner heating thermal shock unit and a photograph of heating during thermal shock test system are shown in Figure 6 [94-95].

In the experiments carried out in burner- thermal shock test unit, coated surfaces of the samples, are heated while the bare surfaces are cooled with pressured air. Oxygen \ natural gas and propane are used as combustive gases. Forming a heat gradient in the sample is aimed and generally for gas turbine practices these types of systems are optimised. The samples that are used in the experiments are generally disc shaped and have a thickness between 2.5-3.0 mm. In burner-thermal shock test system, surface temperature of the specimens are measured by pyrometer, while temperature variation of the substrate material is measured via a thermocouple that passes through centre. Surface temperatures of the coated side of the sample change between 1200 and 1500 °C in accordance with a typical coated turbine component. In literature, thermal cycle durations generally consist of 5 minutes heating and 2 minutes cooling periods. Thermal cycle life of coatings change according to testing temperature and waiting time. Failure criteria in the tests, are based on visual inspection of the coating surface for damages or loss of the coating. In general, a total surface area of coating loss ranging from 10 to 20% is considered as the criterion for failure. The failure mechanisms effective in this system is mainly related to TGO growth at low temperatures and occur at TBC surface at temperatures above 1300°C [16-19,94-99].

The other test method used in evaluation of thermal cycle/shock properties of TBCs is the furnace cycle test. Furnace cycle test better reflect the actual engine conditions. Because this process not only causes cyclic stresses in TBC, but also give rise to a degradation of the bond coating as a result of severe oxidation. In test conditions, as a result of prolonged exposure of TBC to high temperature, oxidation of bond coating takes place. In addition, design limit and performances such as complete failure and depletion of bond coating can be observed with the furnace cycle tests. In this test system, TBC samples generally are subjected to the oxidation between the temperature range 1000-1200 °C, then subjected to cyclic cooling at room temperature. Thermal changes that occur in TBC, take place during the heating and cooling processes. Heating of the system is carried out in the furnace while air-cooling is implemented with the aid of a compressor or fan. A cycle for aircraft engine component consists of 1 hour period, 45-50 minutes of which is in elevated temperatures and 10-15 minutes is spent for cooling. However for industrial gas turbine applications, in order to extend the duration of exposure to high temperature, cycles of 24 hours is typically used and a period of 23 hours of the cycle takes place in elevated temperatures (1080 °C - 1135 °C) while a period of 1 hour is spared for cooling at room temperature. The samples used in these tests are usually disc shaped and of 25.4 mm diameter and criterion for failure is again 10-20% spallation of coated surface. [18,100-102]. FCT test setup for TBC system characterisation can be seen in Figure 7. [18].

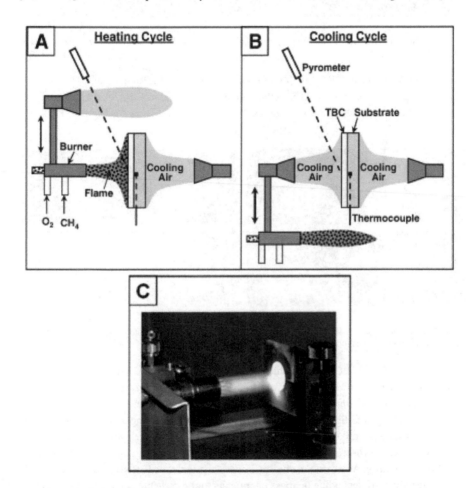

Figure 6. An illustration of thermal shock test device; a) schematic diagram that shows the system heating cycle; b) schematic diagram that shows the system cooling cycle; c) heating cycle photograph of a standard test specimen in thermal cycle/shock equipment [94-95].

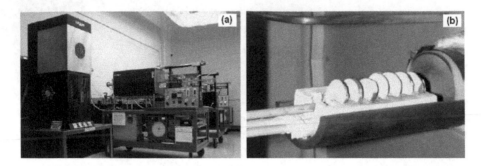

Figure 7. Furnace cycle test system; (a) FCT setup for TBC/bond coat system (b) Samples and sample holder representing the system [18].

In a study by Vaßen et al., NiCOCrAlY bond coats produced by VPS method and TBC systems with YSZ top coat produced by APS method are exposed to thermal cycle and furnace tests. Failure on the coating of the samples as a result of these tests are shown in Figure 8 [103].

Figure 8. Macro images of TBCs produced with different porosity and micro cracks contents after thermal cycle/shock and furnace test; (a)-(d) After burner thermal shock test, (g)-(h) After furnace cycle test [103].

In this study, different TBC systems which are standard, with a high density of micro-cracks, with a content of thick and low porosity and high porosity and with segmentation cracks were investigated by being subjected to burner heating thermal shock and furnace cycle test. In burner thermal shock testing, 5 minutes heating and 2 minutes cooling regimen was used and the sample's surface temperature was kept at 1250°C. The furnace cycle test was conducted at 1100°C with 24 hours heating period at furnace and 1 hour cooling period outside the furnace at room temperature. As a result of studies, it was observed that TBC systems' thermal cycle lives have decreased depending on these parameters in the cycle life of coatings as a result of adverse influence such as TGO thickness which happens and increases on the coating interface, rising temperature of coating surface, sintering effects, stresses resulting from the mixed oxide coatings [103]. In TBC systems, after the oxidation and thermal cycle/shock test, distortion occurs depending on the type of the formation of damage on the sample's surface.

In aerospace applications, the other test method used to determine the thermal shock features of TBCs is JETS method. JETS test is very suitable to provide data on the performance rating on the ceramic itself, but as it does not damage bond coat it does not distinguish errors related to bond coat very well. As a result of high temperature gradients in the ceramic layer with JETS test, weak spots in the ceramic interfaces can be revealed [18]. In Figure 9, a JETS test set up which is used for characterisation of TBC systems can be observed[68].

By creating a large temperature gradient over the TBCs with JETS test, surface temperature rise up to 1400°C and as a result of high temperature, sintering effect act on ceramic top coat. In this test, the main stresses occur thermo-mechanically at the interface of the ceramic and bond coat.

Due to the high temperature gradient within the ceramic layer, TBC/BC interface is oxidised at a small rate. This test is quite fast and the results can be achieved avaregely in 2 days. Similar to the other thermal cycle/shock test, the sample geometry is also disc shaped and has a diameter of 25.4 mm. Right after the heating starts; the samples are cooled by a jet of nitrogen. Nitrogen jet provides the maximum accessible temperature gradient during cooling. In this test, a typical cycle consists of 20 seconds of heating period, 20 seconds of cooling period with nitrogen gas and 40 seconds of waiting period in open atmosphere. [18,68].

Figure 9. Wide JETS setup with four heating and cooling station [68].

4. Summary

In the literature, there are many studies, which are carried out in different cycle tests (with different heating, cooling-holding periods) in different environments (air, water) and systems (heating with burners, furnace cycle tests, JETS test etc.) to determine the thermal cycle/shock behaviour of TBCs. Scientific and industrial institutions continue research, development and studies by simulating real thermal conditions in engines, for investigation of the failure mechanisms and TGO growth behaviours. In this study, TBC systems are introduced and thermal cycle/shock behaviour of TBCs under service conditions and the thermal cycle/shock tests used for evaluation of TBC systems are explained.

Acknowledgements

This work partially supported by The Scientific and Technological Research Council of Turkey (TUBITAK, 111M265).

Author details

Abdullah Cahit Karaoglanli[1], Kazuhiro Ogawa[2], Ahmet Türk[3] and Ismail Ozdemir[4]

1 Dept. of Metalurgical and Materials Eng., Bartin University,Bartin, Turkey

2 Fracture & Reliability Research Institute, Tohoku University, Sendai, Japan

3 Dept. of Metalurgical and Materials Eng., Sakarya University,Sakarya, Turkey

4 Dept. Of Materials Science and Eng., Izmir Katip Celebi University, Izmir, Turkey

References

[1] Mobarra R., Jafari A.H., Karaminezhaad M., Hot corrosion behavior of MCrAlY coatings on In 738LC, Surf.Coat.Tech., 2006, 201, 2202–2207.

[2] Nijdam T.J., Sloof W.G., Combined pre-annealing and pre-oxidation treatment for the processing of thermal barrier coatings on NiCoCrAlY coatings, Surf.Coat.Tech., 2006. 201, 3894–3900.

[3] Li Y., Li C-J., Zhang Q., Xing K., Yang G-J., Effect of surface morphology of MCrAlY bond coats on isothermal oxidation behavior, International Thermal Spray Conference, ITSC 2010, DVS-ASM, Raffles City, Singapore, 2010, 491-497.

[4] Xie D., Xiong Y., Wang F., Effect of an enamel coating on the oxidation and hot corro-
     sion behavior of an HVOF sprayed CoNiCrAlY coatings, Oxidation of Metals, 2003,
     59, 503-516.

[5] Yuan F.H., Chen Z.X., Huang Z.W., Wang Z.G., Zhu S.J., Oxidation behavior of ther-
     mal barrier coatings with HVOF and detonation sprayed NiCoCrAlY bond coats,
     Corrosion Science, 2008, 50,1608–1617.

[6] Schloesser J., Baker M., Rosler J., Thermal barrier coatings for rocket engines, Interna-
     tional Thermal Spray Conference, ITSC 2011, DVS, Hamburg, Germany, 2011,
     952-955.

[7] Lampke T., El-Mahallawy N., Grund T., El-Araby I., Karaoglanli A.C., Effect of bond
     coat material and heat treatment on adhesion strength and characteristics of Thermal
     Barrier Coating system with CGDS, HVOF and APS techniques, International Ther-
     mal Spray Conference, ITSC 2011, DVS, Hamburg, Germany, 2011, 956-959.

[8] Arai M., Suidzu T., Development of porous ceramic coating for high-efficiency cool-
     ing system, International Thermal Spray Conference, ITSC 2011, DVS, Hamburg,
     Germany, 2011, 960-965.

[9] Takahashi S., Hirano N., Kojima Y., Harada Y., Kawasaki A., Ono F., Thermal shock
     resistance of plasma-sprayed Thermal Barrier Coatings, International Thermal Spray
     Conference, ITSC 2011, DVS, Hamburg, Germany, 2011, 1025-1029.

[10] Li C.J., Li Y., Yang G.J., Li C.X., A novel plasma-sprayed durable thermal barrier
     coating with the well-bonded YSZ interlayer between porous YSZ and bond coat, In-
     ternational Thermal Spray Conference, ITSC 2011, DVS, Hamburg, Germany, 2011,
     1234-1240.

[11] Karaoglanli A.C., Lampke T., Grund T., Ak Azem F., Ozdemir I., Turk A., Ustel F.,
     Study of oxidation behavior of TBCs with APS and HVOF CoNiCrAlY bond coat-
     ings, International Thermal Spray Conference, ITSC 2011, DVS, Hamburg, Germany,
     2011, 942-947.

[12] Evans A.G., Mumma D.R., Hutchinson J.W., Meierc G.H., Pettit F.S., Mechanisms
     controlling the durability of thermal barrier coatings, Progress in Materials Science,
     2001, 46 (5), 505–55.

[13] Sridharana S., Xiea L., Jordan E.H., Gella M., Murphy K.S., Damage evolution in an
     electron beam physical vapor deposited thermal barrier coating as a function of cycle
     temperature and time, Materials Science and Engineering A, 2005, 393 (1-2), 51- 62.

[14] Bernier J., Evolution and Characterization of Partially Stabilized Zirconia (7wt%
     Y_2O_3), Thermal Barrier Coatings Deposited by Electron Beam Physical Vapor Deposi-
     tion, Master Thesis, Worcester Polytechnic Institute, May 18, 2001.

[15] Limarga A.M., Clarke D.R., Characterization of Electron Beam Physical Vapor De-
      posited Thermal Barrier Coatings Using Diffuse Optical Reflectance, Int. J. Appl. Ce-
      ram. Technol., 2009, 6 (3), 400–409.

[16] Vaßen R., CERNUSCHI F., RIZZI G., SCRIVANI A., MARKOCSAN N., OSTERG-
      REN L., KLOOSTERMAN A., MEVREL R., FEIST, J., NICHOLLS J., Recent Activities
      in the Field of Thermal Barrier Coatings Including Burner Rig Testing in the Europe-
      an Union, Advanced Engineering Materials, 2008, 10, No. 10, 907-921.

[17] Kara,I, E., TBK Kaplamaların Termal Şok Özelliklerinin İncelenmesi, Master Thesis,
      Sakarya University, Metalurgical and Materials Engineering Department, (in Turk-
      ish), 2008.

[18] Bolcavage A., Feuerstein A., Foster J., Moore P., Thermal Shock Testing of Thermal
      Barrier Coating/Bondcoat Systems, Journal of Materials Engineering and Perform-
      ance, 2004, Vol: 13(4), 389-397.

[19] Traeger F., Vaßen, R., Rauwald K.H., Stover D.., Thermal Cycling Setup for Testing
      Thermal Barrier Coatings, Advanced Engineering Materials, 2003, Vol: 5 (6), 429-432.

[20] Thompson J.A., Clyne T.W., The Effect Of Heat Treatment On The Stiffness Of Zirco-
      nia Top Coats in Plasma Sprayed Zirconia Top Coats in Plasma-Sprayed TBCs, Acta
      mater., 2001, 49, 1565–1575.

[21] Feuerstein A., Knapp J., Taylor T., Ashary V, Bolcavage A., Hitchman N., Technical
      and Economical Aspects of Current Thermal Barrier Coating Systems for Gas Tur-
      bine Engines by Thermal Spray and EB-PVD: A Review, Journal of Thermal Spray
      Technology, 2008, Vol: 17, 199-213.

[22] Bose S., High Temperature Coatings, Butterworth-Heinemann, Elsevier,
      ISBN-13:978-0-7506-8252-7, 2007.

[23] Koolloos M.F.J., Behaviour of low porosity microcracked thermal barrier coatings un-
      der Thermal Loading, PHd. Thesis, Technische Universiteit Eindhoven, NLR, Ned-
      herland, March, 1-168, 2001.

[24] Hass D.D., Thermal Barrier Coatings via Directed Vapor Deposition, Department of
      Materials Science and Engineering, Vol. PhD. Charlottesville, VA: University of Vir-
      ginia, 1-256, 2001.

[25] Jones R.L, Some aspects of the hot corrosion of thermal barrier coatings, Journal of
      Thermal Spray Technology, 1997, 6 (1), 77–84.

[26] Patrick R., Development of Conventional and Nanocrystalline Bond Coats by Cold
      Gas Dynamic Spraying for Aerospace Thermal Barrier Coatings, PhD Theses, Uni-
      versity of Ottowas, Ottowa Ontario Canada, 1-227, July 19, 2010.

[27] ASM Handbook: Volume 5: Surface Engineering (ASM Handbook), ISBN-13:
      978-0871703842, December 1, 1994.

[28] Chen W.R., Wu X., Marple B.R., Nagy D.R., Patnaik P.C., TGO growth behaviour in TBCs with APS and HVOF bond coats, Surf.Coat.Tech., 202, 2677-2683, 2008.

[29] Donachie M.J., Donachie S.J., Superalloys, A Technical Guide, Second Edition, ASM International, ISBN: 0-87170-749-7, 2002.

[30] Saeidi S., Microstructure, Oxidation & Mechanical Properties of As-sprayed and Annealed HVOF & VPS CoNiCrAlY Coatings, PhD Thesis, University of Nottingham, December 2010.

[31] Roberts T., The Structure and Stability of High Temperature Intermetallic Phases for Application within Coating Systems, PhD Thesis, Cranfield University, November 2009.

[32] Eriksson R., High-temperature degradation of plasma sprayed thermal barrier coating systems, PhD thesis, Linköping University, April 2011, ISBN 978-91-7393-165-6, Linköping Sweden, April 25 2011.

[33] Yoshiba M., Abe K., Aranami T., Harada Y., High-Temperature Oxidation and Hot Corrosion Behavior of Two Kinds of Thermal Barrier Coating Systems for Advanced Gas Turbines, Journal of Thermal Spray Technology, 1996, Vol 5 (No. 3), 259-268.

[34] Taylor T.A., Walsh P.N., Thermal expansion of MCrAlY alloys, Surf.Coat.Tech., 2004, 177-178, 24–31.

[35] Fox A.C., Clyne T.W., Oxygen Transport by Gas Permeation through the Zirconia Layer in Plasma Sprayed Thermal Barrier Coatings, Surf. Coat. Tech., 2004, 184, 311-321.

[36] Richer P., Yandouzi M., Beauvais L., Jodoin B., Oxidation behaviour of CoNiCrAlY bond coats produced by plasma, HVOF and cold gas dynamic spraying, Surf.Coat.Tech., 2010, 204, 3962–3974.

[37] Li Y., Li C.J., Zhang Q., Yang G.J., Li C.X., Influence of TGO composition on the thermal shock lifetime of thermal barrier coatings with cold-sprayed MCrAlY bond coat, Journal of Thermal Spray Technology, 2010, 19 (1-2), 168-177.

[38] Tang F., Ajdelsztajn L., Kim G.E., Provenzano V., Schoenung J.M., Effects of Surface Oxidation during HVOF Processing on the Primary Stage Oxidation of a CoNiCrAlY Coating, Surf.Coat.Tech. , 2004, 185, 228-233.

[39] Dalkilic, S., Bir Termal Bariyer Kaplama Sisteminin Yorulma Davranışının İncelenmesi, PhD Thesis, Anadolu University,(In Turkish), March 2007.

[40] Koolloos M.F.J., Houben J.M., Behavior of Plasma-Sprayed Thermal Barrier Coatings during Thermal Cycling and the Effect of a Preoxidized NiCrAlY Bond Coat, Journal of Thermal Spray Technology, 2000, Volume 9(1), 49-58.

[41] Cao X.Q., Vassen R., Stoever D., Ceramic materials for thermal barrier coatings, Journal of the European Ceramic Society, 2004, 24, 1-10.

[42] Vanvalzah J.R., Eaton H.E., Cooling Rate Effects on the Tetragonal to Monoclinic Phase Transformation in Aged Plasma-Sprayed Yttria Partially Stabilized Zirconia, Surf.Coat.Tech., 1991, Vol 46, 289-300.

[43] Pawlowski L., The Science and Engineering of Thermal Spray Coatings, John Wiley & Sons, 2nd Edition, 67-165, 2008.

[44] Ballard J.D., Davenport J., Lewis C., Nelson W., Doremus R.H., Schadler L.S., Phase Stability of Thermal Barrier Coatings Made From 8 wt.% Yttria Stabilized Zirconia: A Technical Note, Journal of Thermal Spray Technology, 2003, Vol 12(1), 34-37.

[45] Clyne T.W., Gill S.C., Residual Stresses in Thermal Spray Coatings and their Effect on Interfacial Adhesion- A Review of Recent Work, Journal of Thermal Spray Technology, 1996, Vol:5, 401-418.

[46] Chraska P., Dubsky J., Neufuss K., Pisacka J., Alumina-Base Plasma-Sprayed Materials Part I: Phase Stability of Alumina and Alumina-Chromia, Journal of Thermal Spray Technology, 1997, Vol 6 (3), 320-326.

[47] M. Saremi, A. Afrasiabi, A. Kobayashi, Microstructural analysis of YSZ and YSZ/Al_2O_3 plasma sprayed thermal barrier coatings after high temperature oxidation, Surf.Coat.Tech., 2008, 202, 3233–3238.

[48] Miller R.A., Thermal Barrier Coatings for Aircraft Engines: History and Directions, Journal of Thermal Spray Technology, 1997, Vol 6(1), 35-42.

[49] Levi C.G., Emerging materials and processes for thermal barrier systems, Current Opinion in Solid State and Materials Science, 2004, 8, 77–91.

[50] Erk K.A., Deschaseaux C., Trice R.W., Grain-Boundary Grooving of Plasma-Sprayed Yttria-Stabilized Zirconia Thermal Barrier Coatings, J. American Ceram. Soc., 2006, 89 (5), 1673–1678.

[51] Haynes J.A., Ferber M.K., Porte W.D., Thermal Cycling Behavior of Plasma-Sprayed Thermal Barrier Coatings with Various MCrAlX Bond Coats, Journal of Thermal Spray Technology, 2000, Vol 9(1), 38-48.

[52] Polat A., Sarikaya O., Celik E., Effects of porosity on thermal loadings of functionally graded Y_2O_3-ZrO_2/NiCoCrAlY coatings, Materials and Design, 2002, 23, 641–644.

[53] Vaßen R., Traeger F., Stover D., Correlation Between Spraying Conditions and Microcrack Density and Their Influence on Thermal Cycling Life of Thermal Barrier Coatings, Journal of Thermal Spray Technology, 2004, Vol 13(3), 396-404.

[54] Grunling H.W., Mannsmann W., Plasma sprayed thermal barrier coatings for industrial gas turbines: morphology, Processing and properties, Journal De Physique IV Colloque C7, supplement au Journal de Physique 111, 1993, Vol 3, 903-912.

[55] Siebert B., Funke C., Vaßen R., Stover D., Changes in porosity and Young's Modulus due to sintering of plasma sprayed thermal barrier coatings, Journal of Materials Processing Technology, 1999, 92-93, 217-223.

[56] Guo H.B., Vaßen R., Stover D., Atmospheric plasma sprayed thick thermal barrier coatings with high segmentation crack density, Surf.Coat.Tech., 2004, 186, 353-363.

[57] Guo H., Murakami H., Kuroda S., Thermal Cycling Behavior of Plasma Sprayed Segmented Thermal Barrier Coatings, Materials Transactions, 2006, Vol 47 (2), 306-309.

[58] Karger M., Vaßen R., Stover D., Atmospheric plasma sprayed thermal barrier coatings with high segmentation crack densities: Spraying process, microstructure and thermal cycling behavior, Surf.Coat.Tech., 206, 16-23, 2011.

[59] Beele, W., Marijnissen G., Liehot A., The evolution of thermal barrier coatings-status and upcoming solutions for today's key issues, Surface and Coatings Technology, 1999, Vol 120-121, 61-67.

[60] Teixeria V., Andritschky M., Gruhn H., Mallener W., Buchkremer H.P., Stover D., Failure of Physical Vapor Deposition/Plasma-Sprayed Thermal Barrier Coatings during Thermal Cycling, Journal of Thermal Spray Technology, 2000, Vol 9(2), 191-197.

[61] Hengbei Z., Zhuo Y., Haydn N.G. Wadley, The influence of coating compliance on the delamination of thermal barrier coatings, Surf.Coat.Tech., 2010, 204, 2432–2441.

[62] Suzuki K., Shobu T., Tanaka K., Residual Stresses of EB-PVD thermal barrier coatings exposed to high temperature, International Centre for Diffraction Data, ISSN 1097-0002, 2009, 537-544.

[63] Kumar S., Cocks A.C.F., Sintering and mud cracking in EB-PVD thermal barrier coatings, Journal of the Mechanics and Physics of Solids, 2012, Vol 60 (4), 723-749.

[64] Guo H., Gong S., Khor K.A., XU H., Effect of thermal exposure on the microstructure and properties of EB-PVD gradient thermal barrier coatings, Surf.Coat.Tech., 2003, 168, 23–29.

[65] Anderson P.S., Wang X., Xiao P., Impedance spectroscopy study of plasma sprayed and EB-PVD thermal barrier coatings, Surf.Coat.Tech., 2004, 185-1, 106-119.

[66] Nicholls J.R., Lawson K.J., Johnstone A., Rickerby D.S., Methods to reduce the thermal conductivity of EB-PVD TBCs, Surf.Coat.Tech., 2002, 151-152, 383-391.

[67] Liu J., Effects of bond coat surface preparation on thermal cycling lifetime and failure characteristic of thermal barrier coatings, Master of Science Thesis, Department of Mechanical, Materials, and Aerospace Engineering, University of Central Florida, Summer Term, 2004.

[68] Helminiak M.A., Factors Affecting The Lifetime Of Thick Air Plasma Sprayed Thermal Barrier Coatings, Master Thesis, University of Pittsburgh, March 18, 2010.

[69] Postolenko V., Failure Mechanisms of Thermal Barrier Coatings for High Tempera-ture Gas Turbine Components under Cyclic Thermal Loading, PhD Thesis, Von der Fakultät für Georessourcen und Materialtechnik der Rheinisch -Westfälischen Tech-nischen Hochschule Aachen, November, 2008.

[70] Mumm D.R., Watanabe M., Evans A.G., Pfaendtner J.A., The influence of test method on failure mechanisms and durability of a thermal barrier system, Acta Materialia, 2004, 52, 1123–1131.

[71] Han J.C., Wang B.L., Thermal shock resistance of ceramics with temperature-depend-ent material properties at elevated temperature, Acta Materiala, 2011, 59, 1373-1382.

[72] Girolama G.D., Marra F., Blasi C., Serra E., Valente T., Microstructure, mechanical properties and thermal shock resistance of plasma sprayed nanostructured zirconia coatings, Ceramics International, 2011, 37, 2711-2717.

[73] Fry A.T., Banks P.J., Nunn J., Brown J.L., Comparison of the Thermal Cycling Per-formance of Thermal Barrier Coatings under Isothermal and Heat Flux Conditions, Materials Science Forum, 2008, Vol 595-598, 77-85.

[74] Balint D.S., Hutchinson J.W., An analytical model of rumplingin thermal barrier coat-ings Journal of the Mechanics and Physics of solids, 2005, 53, 949–973.

[75] Yanar N.M., The failure of therma barrier coatings at elevated temperatures, PhD thesis, University of Pittsburgh, 2004.

[76] Li M., Sun X., Hu W., Guan H., Thermal shock behavior of EB-PVD thermal barrier coatings, Surf.Coat.Tech., 2007, 201, 7387–7391.

[77] Khan A.N., Lu J., Behavior of air plasma sprayed thermal barrier coatings, subject to intense thermal cycling, Surf.Coat.Tech., 2003, 166, 37–43.

[78] Verbeek A.T.J., Plasma sprayed thermal barrier coatings: Production, characteriza-tion and testing, PhD Thesis, Technische Universiteit Eindhoven,1992.

[79] Aygun A., Novel Thermal Barrier Coatings(TBCs) that are resistant to high tempera-ture attack by CaO-MgO-Al$_2$O$_3$-SiO$_2$ (CMAS) glassy deposits, PhD thesis, The Ohio State University, 2008.

[80] Evans A.G., Fleck N.A., Faulhaber S., Vermaak N., Maloney M., Daroli R., Scaling laws governing the erosion and impact resistance of thermal barrier coatings, Wear, 2006, 260, 886–894.

[81] Wellman R.G., Nicholls J.R., Erosion, corrosion and erosion–corrosion of EB PVD thermal barrier coatings, Tribology International, 2008, 41 (7), 657-662.

[82] Mao W.G., Zhou Y.C., Yang L., Yu X.H., Modeling of residual stresses variation with thermal cycling in thermal barrier coatings, Mechanics of Materials, 2006, 38, 1118-1127.

[83] Vaßen R., Kerkhoff G., Stover D., Development of a micromechanical life prediction model for plasma sprayed thermal barrier coatings, Materials Science and Engineering A, 2001, 303, 100–109.

[84] Yanar N.M., Helminiak M., Meier G.H., Pettit F.S., Comparison of the Failures during Cyclic Oxidation of Yttria-Stabilized (7 to 8 Weight Percent) Zirconia Thermal Barrier Coatings Fabricated via Electron Beam Physical Vapor Deposition and Air Plasma Spray, Metallurgical and Materials Transactions A, 2011, Vol 42, 905-921.

[85] Guo H., Murakami H., Kuroda S., Effects of Heat Treatment on Microstructures and Physical Properties of Segmented Thermal Barrier Coatings Materials Transactions, 2005, Vol. 46, No. 8, 1775-1778.

[86] Wright P.K., Evans A.G., Mechanisms governing the performance of thermal barrier coatings, Current Opinion in Solid State and Materials Science, 1999, 4, 255-265.

[87] Anderson A., An Investigation Of the Thermal Shock Behavior Of Thermal Barrier Coatings, International Journal of Engineering Science and Technology, 2011, Vol 3 (11), 8154-8158.

[88] Kim J.H., Kim M.C., Park C.G., Evaluation of functionally graded thermal barrier coatings fabricated by detonation gun spray technique, Surf.Coat.Tech., 2003, 168, 275–280.

[89] Murakami H., Tetsuro Y., Satoshi S., Process Dependence of Ir-Based Bond Coatings, Materials Transactions, 2004, Vol 45 (9), 2886-2890.

[90] Peng H., Guo H., He J., Gong S., Cyclic oxidation and diffusion barrier behaviors of oxides dispersed NiCoCrAlY Coatings, Journal of Alloys and Compounds, 2010, 502, 411–416.

[91] Bacos M.P., Dorvaux J.M., Lavigne O., Mevrel R., Poulain M., Rio C., Vidal Setif M.H., Performance and Degradation Mechanisms of Thermal Barrier Coatings for Turbine Blades: a Review of Onera Activities, Journal Aerospace Lab, 2011, Issue 3, 1-11.

[92] Mohammadi M., Javadpour S., Kobayashi A., Jahromi S.A.J., Shirvani K., Thermal shock properties and microstructure investigation of LVPS and HVOF-CoNiCrAlYSi coatings on the IN738LC superalloy, Vacuum, 2012, doi:10.1016/j.vacuum.2012.02.003, 1-6.

[93] Wang Q.M., Tang Y.J., Guo M.H., Ke P.L., Gong J., Sun C., Wen L.S., Thermal shock cycling behavior of NiCoCrAlYSiB coatings on Ni-base superalloys I. Accelerated oxidation attack, Materials Science and Engineering A, 2005, 406, 337–349.

[94] Steinke T., Sebold D., Mack D.E., Vaßen R., Stover D., A novel test approach for plasma-sprayed coatings tested simultaneously under CMAS and thermal gradient cycling conditions, Surf.Coat.Tech., 2010, 205 7, 2287-2295.

[95]   Drexler J.M., Aygun A., Li D., Vaßen R., Steinke T., Padture N.P., Thermal-gradient testing of thermal barrier coatings under simultaneous attack by molten glassy deposits and its mitigation, Surf.Coat.Tech., 2010, 204, 2683–2688.

[96]   Braue W., Mechnich P., Recession of an EB-PVD YSZ Coated Turbine Blade by CaSO$_4$ and Fe, Ti-Rich CMAS-Type Deposits, J. Am. Ceram. Soc., 2011, 94 (12), 4483–4489.

[97]   Tong C., Ji-Jie W., Ren-Guo G., Li-Qing C., Gumming Q., Microstructures and Properties of Thermal Barrier Coatings Plasma-Sprayed by Nanostructured Zirconia, Journal of Iron and Steel Research, International, 2007, Vol 12 (5), 116-120.

[98]   Wu J., Guo H.B., Zhou L., Wang L., Gong S.K., Microstructure and Thermal Properties of Plasma Sprayed Thermal Barrier Coatings from Nanostructured YSZ, Journal of Thermal Spray Technology, 2010, Vol 19 (6), 1186-1194.

[99]   Ma W., Dong H., Guo H., Gong S., Zheng X., Thermal cycling behavior of La$_2$Ce$_2$O$_7$/8YSZ double-ceramic-layer thermal barrier coatings prepared by atmospheric plasma spraying, Surf.Coat.Tech., 2010, 204, 3366–3370.

[100] Spitsberg I.T., Mumm D.R., Evans A.G., On the failure mechanisms of thermal barrier coatings with diffusion aluminide bond coating, Materials Science and Engineering:A, 2005, Vol 394 (1-2), 176-191.

[101] Tolpygo V.K., Clarke D.R., Morphological evolution of thermal barrier coatings induced by cyclic oxidation, Surf.Coat.Tech., 2003, 163-164, 81–86.

[102] Ruud J.A., Bartz A., Borom M.P., Johnson C.A., Strength Degradation and Failure Mechanisms of Electron-Beam Physical-Vapor-Deposited Thermal Barrier Coatings, J. Am. Ceram. Soc., 2001, 84 (7), 1545–52.

[103] Vaßen R., GIESEN S., STOVER D., Lifetime of Plasma-Sprayed Thermal Barrier Coatings: Comparison of Numerical and Experimental Results, Journal of Thermal Spray Technology, 2009, Vol 18(5-6), 835-845.

Permissions

The contributors of this book come from diverse backgrounds, making this book a truly international effort. This book will bring forth new frontiers with its revolutionizing research information and detailed analysis of the nascent developments around the world.

We would like to thank Dr. Rakesh Sehgal, for lending his expertise to make the book truly unique. He has played a crucial role in the development of this book. Without his invaluable contribution this book wouldn't have been possible. He has made vital efforts to compile up to date information on the varied aspects of this subject to make this book a valuable addition to the collection of many professionals and students.

This book was conceptualized with the vision of imparting up-to-date information and advanced data in this field. To ensure the same, a matchless editorial board was set up. Every individual on the board went through rigorous rounds of assessment to prove their worth. After which they invested a large part of their time researching and compiling the most relevant data for our readers. Conferences and sessions were held from time to time between the editorial board and the contributing authors to present the data in the most comprehensible form. The editorial team has worked tirelessly to provide valuable and valid information to help people across the globe.

Every chapter published in this book has been scrutinized by our experts. Their significance has been extensively debated. The topics covered herein carry significant findings which will fuel the growth of the discipline. They may even be implemented as practical applications or may be referred to as a beginning point for another development. Chapters in this book were first published by InTech; hereby published with permission under the Creative Commons Attribution License or equivalent.

The editorial board has been involved in producing this book since its inception. They have spent rigorous hours researching and exploring the diverse topics which have resulted in the successful publishing of this book. They have passed on their knowledge of decades through this book. To expedite this challenging task, the publisher supported the team at every step. A small team of assistant editors was also appointed to further simplify the editing procedure and attain best results for the readers.

Our editorial team has been hand-picked from every corner of the world. Their multi-ethnicity adds dynamic inputs to the discussions which result in innovative outcomes. These outcomes are then further discussed with the researchers and contributors who give their valuable feedback and opinion regarding the same. The feedback is then collaborated with the researches and they are edited in a comprehensive manner to aid the understanding of the subject.

Apart from the editorial board, the designing team has also invested a significant amount of their time in understanding the subject and creating the most relevant covers. They scrutinized every image to scout for the most suitable representation of the subject and create an appropriate cover for the book.

The publishing team has been involved in this book since its early stages. They were actively engaged in every process, be it collecting the data, connecting with the contributors or procuring relevant information. The team has been an ardent support to the editorial, designing and production team. Their endless efforts to recruit the best for this project, has resulted in the accomplishment of this book. They are a veteran in the field of academics and their pool of knowledge is as vast as their experience in printing. Their expertise and guidance has proved useful at every step. Their uncompromising quality standards have made this book an exceptional effort. Their encouragement from time to time has been an inspiration for everyone.

The publisher and the editorial board hope that this book will prove to be a valuable piece of knowledge for researchers, students, practitioners and scholars across the globe.

List of Contributors

Ioannis Templalexis
Hellenic Air Force Academy Department of Aeronautical Sciences Section of Thermodynamics,
Propulsion and Power Systems, Greece

Marco Antônio Rosa do Nascimento, Lucilene de Oliveira Rodrigues, Eraldo Cruz dos Santos, Eli Eber Batista Gomes, Fagner Luis Goulart Dias, Elkin Iván Gutiérrez Velásques and Rubén Alexis Miranda Carrillo
Federal University of Itajubá – UNIFEI, Brazil

Konstantinos G. Kyprianidis and Andrew M. Rolt
Rolls-Royce plc, UK

Vishal Sethi
Cranfield University, UK

Ene Barbu, Romulus Petcu, Valeriu Vilag, Valentin Silivestru, Jeni Popescu, Cleopatra Cuciumita and Sorin Tomescu
National Research and Development Institute for Gas Turbines COMOTI, Bucharest, Romania

Tudor Prisecaru˘
Politehnica University, Bucharest, Romania

Takeharu Hasegawa
Central Research Institute of Electric Power Industry, Nagasaka, Yokosuka-Shi Kanagawa-Ken, Japan

Zanger Jan, Monz Thomas and Aigner Manfred
German Aerospace Center, Institute of Combustion Technology, Stuttgart, Germany

M. Khosravy el Hossaini
Research Institute of Petroleum Industry, Iran

G. G. Kulikov, V. Yu. Arkov and A.I. Abdulnagimov
Automated Control and Management Systems Department, Ufa State Aviation Technical University, Ufa, Russia

Anastassios G. Stamatis
Mechanical Engineering Department, Polytechnic School, University of Thessaly, Volos, Greece

Abdullah Cahit Karaoglanli
Department of Metalurgical and Materials Eng., Bartin University,Bartin, Turkey

Kazuhiro Ogawa
Fracture & Reliability Research Institute, Tohoku University, Sendai, Japan

Ahmet Türk
Department of Metalurgical and Materials Eng., Sakarya University,Sakarya, Turkey

Ismail Ozdemir
Department Of Materials Science and Eng., Izmir Katip Celebi University, Izmir, Turkey

Printed in the USA
CPSIA information can be obtained
at www.ICGtesting.com
JSHW011444221024
72173JS00004B/934